GATEWAY TO CONDENSED MATTER PHYSICS AND MOLECULAR BIOPHYSICS

Concepts and Theoretical Perspectives

GATEWAY TO CONDENSED MATTER PHYSICS AND MOLECULAR BIOPHYSICS

Concepts and Theoretical Perspectives

by

Ranjan Chaudhury, PhD

A‍AP | APPLE
ACADEMIC
PRESS

First edition published 2022

Apple Academic Press Inc.
1265 Goldenrod Circle, NE,
Palm Bay, FL 32905 USA

4164 Lakeshore Road, Burlington,
ON, L7L 1A4 Canada

CRC Press
6000 Broken Sound Parkway NW,
Suite 300, Boca Raton, FL 33487-2742 USA

2 Park Square, Milton Park,
Abingdon, Oxon, OX14 4RN UK

© 2022 Apple Academic Press, Inc.

Apple Academic Press exclusively co-publishes with CRC Press, an imprint of Taylor & Francis Group, LLC

Library and Archives Canada Cataloguing in Publication

Title: Gateway to condensed matter physics and molecular biophysics : concepts and theoretical perspectives / by Ranjan Chaudhury, PhD.

Names: Chaudhury, Ranjan, author.

Description: First edition. | Includes bibliographical references and index.

Identifiers: Canadiana (print) 20210188766 | Canadiana (ebook) 20210188812 | ISBN 9781771889131 (hardcover) | ISBN 9781774638057 (softcover) | ISBN 9781003057840 (ebook)

Subjects: LCSH: Condensed matter. | LCSH: Physics. | LCSH: Molecular biology.

Classification: LCC QC173.454 .C53 2021 | DDC 530.4/1—dc23

Library of Congress Cataloging-in-Publication Data

Names: Chaudhury, Ranjan, author.

Title: Gateway to condensed matter physics and molecular biophysics : concepts and theoretical perspectives / by Ranjan Chaudhury.

Description: First edition. | Palm Bay, FL : Apple Academic Press, 2021. | Includes bibliographical references and index. | Summary: "This new book, Gateway to Condensed Matter Physics and Molecular Biophysics: Concepts and Theoretical Perspectives, provides the necessary background material and brings into focus the fundamental concepts essential for advanced research in theoretical condensed matter physics and its interface with molecular biophysics. It is the outcome of the author's long teaching and research career in theoretical condensed matter physics and related interdisciplinary fields. The author aims to motivate students to take up research in condensed matter physics and march toward new frontiers. He writes: "My long understanding of students' attitude and orientation brings me to the conclusion that many of them are quite excited about the developments in the frontier research areas at the beginning of their career; however, a sizable fraction of them start losing interest gradually as they are often unable to connect these developments with the basic physics they have studied. I have tried to fill this gap in this book." To this end special care has been taken to balance the physical concepts and mathematical expressions as well as proper mixing of theoretical and experimental aspects. He starts with the very well known elementary ideas or basic concepts and goes forward so as to remove the apparent conceptual and technical gap between the known laws and various interesting, challenging, and novel experimental results and effects, some of which are amongst the latest discoveries. Key features: Introduces a new way of looking at various important and fundamental phenomena in condensed matter from the perspective of microscopic theory Explores a new interface of quantum condensed matter physics and molecular biophysics, highlighting research potentialities Addresses the crucial questions surrounding these phenomena when they are mutually coexisting or competing in real condensed matter systems or materials, from both theoretical and experimental angles Deals with biological molecules and some of their properties and processes and discusses the modeling of these with the help of condensed matter physics and statistical physics Emphasizes fundamental concepts, particularly in condensed matter physics and making proper use of them The book is unique because it provides an insight into the microscopics of some of the fundamental phenomena in condensed matter physics and their interrelations. It also attempts to bridge the apparently unconnected fields of molecular biophysics and condensed matter physics. Gateway to Condensed Matter Physics and Molecular Biophysics: Concepts and Theoretical Perspectives will be an invaluable advanced reference book for master's students (in physics) and PhD students (in condensed matter physics) in their early years of research, and will also be helpful to faculty and researchers"-- Provided by publisher.

Identifiers: LCCN 2021016789 (print) | LCCN 2021016790 (ebook) | ISBN 9781771889131 (hardcover) | ISBN 9781774638057 (paperback) | ISBN 9781003057840 (ebook)

Subjects: LCSH: Condensed matter. | Molecular biology.

Classification: LCC QC173.454 .C498 2021 (print) | LCC QC173.454 (ebook) | DDC 530.4/1--dc23

LC record available at https://lccn.loc.gov/2021016789

LC ebook record available at https://lccn.loc.gov/2021016790

ISBN: 978-1-77188-913-1 (hbk)
ISBN: 978-1-77463-805-7 (pbk)
ISBN: 978-1-00305-784-0 (ebk)

Dedication

Dedicated to the Memory of
My Beloved Father.

About the Author

Ranjan Chaudhury, PhD
*Superannuated Full Professor, Department of Condensed Matter Physics
and Material Sciences, S.N. Bose National Centre For Basic Sciences,
India; Full Professor Ramakrishna Mission Vivekananda Educational
and Research Institute, India (till 31.01.21)*

Ranjan Chaudhury, PhD, has been a professor at S.N. Bose National Centre, Kolkata, India, since 1994 and is at present attached there as a Superannuated Full Professor in the Condensed Matter Physics and Material Sciences Department. He was also an Adjunct Professor in the Physics Department at the Ramakrishna Mission Vivekananda Educational and Research Institute (Belur) during 2017–2018 and later was a Full Professor in the department there as well. Professor Chaudhury has done extensive theoretical work in the areas of superconductivity and magnetism in low-dimensional systems and also in the few areas of theoretical molecular biophysics. He has about 80 publications to date, which include publications in international journals as well as research reports and international conference papers. He has been teaching regularly for last more than 20 years and has supervised PhD and master's students for their theses and major projects. He also delivered invited talks at some of the well-known international and national conferences, both abroad and in India. He has been the recipient of several awards both at the national and international levels.

Prof. Chaudhury has been a member of a number of international scientific societies and organizations such, as the American Chemical Society (New York), MSI (Minneapolis), and ATINER (Athems Institute for Education & Research). He has been a recipient of several awards both at the national and at the international levels.

Dr. Chaudhury completed his Master's in Physics at IIT Kharagpur and obtained his PhD from TIFR (Mumbai). During his PhD, he did theoretical work in the areas of superconductivity and magnetism. He was a postdoctoral fellow/collaborating visiting scientist at various institutions abroad including, International Centre for Theoretical Physics (ICTP, (Trieste, Italy); McMaster University (Hamilton, Ontario, Canada); University of Minnesota (Minneapolis); LEPES, CNRS (Grenoble, France); and BLTP, JINR (Dubna, Russia) during 1988–2000 and was involved with theoretical research in various areas of condensed matter physics. In addition, he was a Visiting Professor at African University of Science and Technology (AUST) (Nigeria) during 2009–2010 and taught there in the Theoretical Physics Department as well.

Contents

Abbreviations

AFM	atomic force microscopy
BCS	Bardeen-Cooper-Schrieffer
BEC	Bose-Einstein Condensation
BKT	Berezinskii, Kosterlitz, and Thouless
BPR	base-pairing rule
BV	Bogoliubov-Velatin
CDW	charge density wave
CP	Clauxius-Pearson
CR	complementarity restoration
DM	Dzyaloshinskii-Moriya
DNA	deoxyribonucleic acid
DOS	density of states
e.m.	electromagnetic
EDOS	electronic density of states
ESR	electron spin resonance
GC	Gorter-Casimir
GFP	green fluorescent protein
GL	Ginzburg-Landau
GMR	giant magneto-resistance
GPB	De Gennes, Peyrard-Bishop
HPT	Holstein-Primakoff transformations
JSDC	Jensen-Shannon divergence criterion
JSDF	Jensen-Shannon divergence function
LFL	Landau Fermi liquid
LP	London-Pippard's
MCMD	Monte Carlo-molecular dynamics
MFL	marginal Fermi liquid
MHI	Mott-Hubbard insulator
MOKE	magneto-optic Kerr effect
MW	microwave
NMR	nuclear magnetic resonance
PAC	perturbed angular correlation
QHE	quantum hall effect

RE	repairing efficiency
RKKY	Rudolf-Kittel-Kasua-Yoshida
RPA	random phase approximation
RWM	random walk model
SBB	Scafroth-Blatt-Butler
SDW	spin density wave
SN	superconductor-normal metal
SQUID	superconducting quantum interference device
SSS	statistical scoring scheme
STM	scanning tunneling microscopy
SWCNT	single-walled carbon nanotube
SWT	Schrieffer-Wolf transformation
TBH	tight-binding Hamiltonian
TLL	Tomonaga-Luttinger liquid
TM	tautomeric mutation
TMR	tunneling magneto-resistance
VSM	vibrating sample magnetometer
WS	Wigner solid

Preface

This is an advanced reference book in theoretical condensed matter physics and related interdisciplinary problems. Its target readers are post-graduate (in Physics) and research students (in Condensed Matter Physics) in their early years of research.

This book is an outcome of my long teaching experiences (at S.N. Bose Center, AUST, and RKMVERI) of nearly 20 years at the master's and post-MSc level, combined with my research career (at TIFR, S.N. Bose Center, and various institutions abroad) spanning about 35 years.

The principal aim of this book is to motivate students to take up basic research in condensed matter physics and march towards the frontiers; these frontier areas may not always be fashionable though. Conceptual clarification with clear directions to the relevant textbooks and research papers for details have been attempted systematically and with strong emphasis. Special care has also been taken to balance the physical concepts and mathematical expressions and as well as the proper mixing of theoretical and experimental aspects.

A sincere attempt has been made to start from the very well-known elementary ideas or basic concepts and go forward so as to remove the apparent conceptual and technical gap between the known laws and various interesting, challenging, and novel experimental results and effects, some of which are amongst the latest discoveries.

My long understanding of students' attitude and orientation brings me to the conclusion that many of them are quite excited about the developments in the frontier research areas at the beginning of their careers. However, a sizeable fraction of them start losing interest gradually as they are often unable to connect these developments with the basic physics that they have studied. I have tried to fill this gap sincerely as much as possible in this book.

My other strong feeling developed over years is that teaching and research can actually supplement and strengthen each other very effectively. I have kept it especially in my mind when planning and writing this book.

I very genuinely and sincerely owe the motivation of this book writing to my late beloved father, Prof. Tapas Kumar Chaudhury, who always inspired me and helped me to concentrate and focus on this difficult but noble assignment.

I would also like to express my deep gratitude to all my teachers, PhD thesis supervisors, all my mentors, and research collaborators for drawing my attention to various challenging problems in various areas of condensed matter physics and quantum biology and help me tackling them through collaborations. The list is very long but I would like to specially mention the following names: Prof. S. K. Ghatak, Prof. S. S. Jha, Prof. B. S. Shastry, Prof. M. Barma, Prof. A. K. Grover, Prof. D. Dhar, Prof. C. K. Majumdar, Prof. J. P Carbotte, Prof. J. W. Halley, Prof. B. K. Chakraverty, Prof. T. V. Ramakrishnan, Prof. M. Avignon, Prof. D. Feinberg, Prof. N. Plakida, Prof. M. P. Das, Prof. G. Baskaran, Prof. T. Chatterjee, Prof. V. Golo, Prof. D. Gangopadhyay, Prof. J. Chakrabarti, Prof. S. K. Paul, and Prof. A. Parola, for enriching my knowledge and understanding and sharing exciting science with me.

It would also be quite incomplete and unfair if I don't acknowledge the hidden contributions of my students (past and present). Through teaching and supervision, my own understanding of the intricacies in various scientific concepts has acquired clarity and totality.

Special thanks are due to Suraka Bhattacharjee, one of my PhD students, for helping me with the figures in this book.

Finally, I would like to thank my family members, relatives, and friends who have stood by my side and supported me during the period of my personal tragedies and have helped me to complete the preparation of the manuscript.

It is indeed a pleasure for me to express my gratitude to Ms. Sandra Jones Sickels, Vice President, Apple Academic Press, for her invitation to me to write this book. I also would like to thank Mr. Ashish Kumar, President, Apple Academic Press, and his team members for various technical suggestions regarding the manuscript and cooperation.

Last but not the least, I would also like to gratefully acknowledge all the infrastructural facilities provided by the S.N. Bose Center during the writing of this book.

—Ranjan Chaudhury, PhD

CHAPTER 1

Introduction to Electronic Properties of Solids: Band Theory and Consequences

1.1 GENERAL INTRODUCTION

Condensed matter physics is a branch of physics, which deals with the properties of solids and liquids. This branch is further subdivided into 'hard-core condensed matter physics' and 'soft condensed matter physics.' The hard-core condensed matter studies and analyses the truly microscopic properties of solids and liquids including both conventional and non-conventional types. This contains both ordered and disordered solids, as well as classical and quantum liquids. On the other hand, the soft condensed matter generally looks at the nearly classical modeling of semi-microscopic and even macroscopic aspects of a special class of "soft materials" like granular materials, colloids, spongy materials, etc. In Chapters 1–5, we will only introduce a few key areas falling within the purview of the hard-core condensed matter physics. In these chapters, we will mostly confine our discussions only to crystalline solids. In Chapter 6, we will discuss a few problems at the interface of molecular biophysics and condensed matter physics mostly involving DNA.

This book attempts to address concepts and problems at the level of both foundations and frontiers in condensed matter physics, which are fundamental and stimulating but may not necessarily be fashionable always. The approach followed is based on a theoretical one but supplemented with results from experimental and computational studies. Last but not least, each chapter contains some of the advanced research results in the respective fields as well as the development of the conceptual foundation.

The book is divided into six chapters. In the first chapter, a general introduction to electronic band theory is presented. Moreover, various experimental techniques and phenomena associated with the studies of energy bands are also briefly dealt with. In the second chapter, we cover various aspects of magnetism in solids, and we include both microscopic theories as well as the experimental techniques along with some of the effects observed. Applications to some new techniques and devices are also discussed. The third chapter is devoted to lattice dynamics and the origin of phonons. It also contains some manifestations of various types of phonons to different kinds of phenomena and effects. The fourth chapter contains a detailed analysis of the electrons in real solids. It introduces various many-body techniques and effective theories, which are semi-phenomenological in nature. Besides, a bit of surface physics is also included. Here again, some connections to exotic phenomena and new devices are brought out. The fifth chapter deals exclusively with superconductivity. It carries out a fairly detailed analysis of the microscopic and phenomenological theories of superconductivity, for both conventional materials as well as the exotic materials. Moreover, it covers special topics such as magnetic superconductors and high-temperature superconductivity. It also includes various forms of experimental techniques and their applications. The very last chapter deals with the theoretical attempts from a viewpoint of a quantum condensed matter physicist to understand certain important biophysical properties of DNA, a very crucial ingredient of all living systems.

Almost all the physical properties of solids can be linked to the following three types of very fundamental and important characteristics: (a) electrical transport, (b) magnetic, and (c) electrical polarization. Again based on these three characteristics, the solids may broadly be classified in the following ways:

1. Electrical transport:
   i. insulator;
   ii. semiconductor;
   iii. metal; and
   iv. superconductor.

2. Magnetic:
   i. diamagnet;
   ii. paramagnet;
   iii. ferromagnet; and
   iv. antiferromagnet.

3. Electrical polarization:
    i.   paraelectric;
    ii. ferroelectric; and
    iii. anti-ferroelectric.

Under group (a) classification, an insulator is defined as a material, which does not conduct electricity at all at room temperature or below. A semiconductor is one, which behaves as an insulator at zero temperature but becomes a moderate conductor at finite temperature. A metal is a very good conductor at any temperature. A superconductor is a special kind of metal, metallic alloy, or even a doped insulator which becomes a perfect conductor, i.e., loses all electrical resistance below a certain finite temperature. Besides these four types, there is a special variety of semiconductors called "semi-metal," which behaves like a semiconductor at zero temperature but becomes metallic as soon as the temperature is raised. Typically, in a semiconductor, the electrical conductivity increases with temperature. A metal exhibits the opposite character *viz.* the electrical conductivity falls as temperature rises. Thus, a semiconductor is an insulator at zero temperature. A very pure metal is a perfect conductor only at zero temperature. In contrast, a superconductor becomes a perfect conductor at a temperature above absolute zero and remains in that state down to zero temperature. Some of the examples of the materials belonging to these classes are the various metallic oxides like NiO, various organic polymers like polythene, alkali halides like NaCl (insulator); Si, Ge (semiconductors); Bi (semimetal); Na, K, Fe, Cu, etc. (metal); Hg, Pb, Al, Sn, Nb_3Sn, Nb_3Ge, etc., (superconductor). It should also be remarked however that the insulators belonging to the alkali halide subgroup, exhibit a very feeble conductivity at room temperature due to the ionic migration.

Within the group (b) characterization, in a diamagnet and externally applied magnetic field is opposed by an induced orbital current. Moreover, there are no net intrinsic magnetic moments on the constituent atoms. In a paramagnet even though the individual atoms have magnetic moments, they are randomly oriented in the absence of any external magnetic field—leading to the non-existence of any overall spontaneous moment in the system. However, in both ferromagnet and anti-ferromagnet, there are the existence of well-ordered structures of spontaneous magnetic moments. In a ferromagnet, the magnetic moments on the neighboring lattice sites are oriented parallel; whereas in an antiferromagnetic system, they are anti-aligned.

The group (c) classification is exactly analogous to that of the group (b), with the quantity "magnetic moment" being replaced with "electrical dipole moment."

The comparison between dia-electric (so called !), i.e., both metallic systems and some electrolytic substances with dielectric constant (microscopic dielectric function at finite q) becoming –ve and diamagnetic, i.e., both superconducting and insulating systems with magnetic (orbital) susceptibility at very small q becoming –ve, is very important. Interestingly both of these can occur for a superconductor almost simultaneously; however, one should also distinguish between a microscopic dia-electric function and a macroscopic one.

It is quite interesting to note that there is a general pattern connecting members of these three groups. The insulators are generally diamagnetic or in some cases may show ferromagnetism or anti-ferromagnetism. The metals are paramagnetic mostly; however, some of them show ferromagnetism or even antiferromagnetism/spin density wave (SDW). The superconductors are diamagnets and in fact, exhibit super-diamagnetism. Most of the ferroelectric and anti-ferroelectric substances are insulators; however, some of the metals may also exhibit spontaneous electric polarization under certain conditions. Again, there exist non-superconducting super-diamagnets which are in fact insulators like AgI, CuCl, etc. Moreover very interestingly, sometimes approaching a magnetic ordering or a ferro/anti-ferroelectric ordering (involving charge ordering or structural transition as well), a metal may turn favorably into a superconductor without undergoing the long-range ordering which it was originally heading towards (Chaudhury and Jha, 1984). A magnetic superconductor binary alloy Y_9Co_7 showed this interesting feature. Some of the very novel superconductors with substantial magnetic correlations, in particular the high-temperature Cuprate superconductors and the Fe-based pnictide superconductors, too quite often exhibit this trend (Chaudhury, 2009). On the other hand, the alloys like Nb_3Sn and Nb_3Ge turn into superconductors (the conventional high-temperature ones) at the threshold of a possible structural transition (Ginzburg and Kirzhnits, 1982).

Besides, there are systems known as magnetic superconductors which are mostly alloys of rare earth and transition metal components, which display very rich interplay and sometimes coexistence between superconductivity and long-range magnetic ordering. We will discuss them elaborately in Chapter 2.

The theoretical understanding of these very interesting interplay phenomena described above is often attempted schematically through a simplistic jellium model-based approach. To make it truly relevant to the real condensed matter systems and materials for quantitative comparison and analysis, it is however crucial to bring in the lattice and include the band structure effects in the theoretical treatments. Unfortunately, this is often not done.

It turns out, as we will see later, that the properties (a) and (b) owe their origin to the "electrons" present in the solid. The property (c), on the contrary, mostly arises from the vibrations of the "ionic lattice" belonging to the solid. In certain cases, in metallic/semi-metallic systems electron or hole occupation in multiple electronic bands with a very small energy gap, the electrons may also contribute to the spatial ordering of spontaneous electrical polarization. The property (a) based classification is essentially decided by the electronic energy or excitation spectrum of the system. Furthermore, excepting the case of superconductor, the other three categories within the group (a) can all be discussed within the approximation of the "one-electron theory." For superconductors and for some special kind of insulators (viz. Mott-Hubbard insulator (MHI) and charge transfer insulator) "many-electron theory" is needed explicitly. In property (b)-based categorization, all the types excepting diamagnet, require both "one-electron" as well as "many-electron" treatments for complete theoretical understanding. Diamagnetism is a "universal property" which can always be discussed within the one-electron picture for a solid.

The theoretical understanding of these properties of solids enlisted above is all based on full quantum treatments and often needs the use of statistical mechanics of interacting particles at finite temperature.

In this first chapter, we present introductory but fairly detailed discussions on these electronic properties of solids.

1.2 ORIGIN OF BAND THEORY

In view of the tremendous applicability of the simple one-electron theory, as brought out in the previous section, we first take up the problem of an electron moving through a periodic crystal lattice in a solid. Just now, we are not worried about the nature of the symmetry of the lattice or in other words the type of the Bravis lattice to which the crystal belongs. Our treatment is very general to start with. Since the property of electrons is governed by the laws of quantum mechanics, we will have to make use of Schrodinger's equation to tackle this problem. It is well understood now that a microscopic particle like an electron is described by the De Broglie waves which are quantum waves describing the probability amplitude for the location of the particle, obeying the time-dependent Schrodinger equation. The Hamiltonian appropriate to this problem is given by, neglecting the Coulomb interaction between the electrons for the time being:

$$H = -\frac{h^2}{(2\pi)^2 \, 2m} \nabla^2 + V(r) \tag{1.1}$$

where, the first operator on the right-hand side is the kinetic energy operator corresponding to an electron and the second quantity is the periodic potential experienced by an electron from the positively charged ion cores situated on the lattice sites. Because of the perfect periodic nature of the lattice, we have the translational symmetry property of the potential experienced by the moving electron due to the ions on the lattice:

$$V(r) = V(r+R) \tag{1.2}$$

where, R is any lattice translation vector.

In fact, in general, the full Hamiltonian obeys this translational symmetry as well, as can be seen easily. Thus, we can write:

$$[H, \hat{T}] = 0 \tag{1.3}$$

where, $\hat{T}$ is the 'lattice translational operator' defined by its action on any function, as given below:

$$\hat{T}\phi(r) = \phi((r+R)) \tag{1.4}$$

Now the operators $\hat{T}$ and H being mutually commuting may have simultaneous eigenstates, provided they are non-degenerate. This leads to the following form for the eigenfunctions of the Hamiltonian:

$$\psi_k(r) = u_k(r)\exp(ik \cdot r) \tag{1.5}$$

where, $\psi_k(r)$ is an eigenfunction and $u_k(r)$ is a periodic function on the lattice satisfying:

$$u_k\,(r) = u_k\,(r + R) \qquad\qquad (1.6)$$

The subscript k denotes the wave vector of the traveling wave described by the eigenfunction ψ and it also represents a band state. It can be seen very easily from the above form of $\psi_k(r)$ that it satisfies the following equation as well:

$$\psi_k\,(r + R) = \exp\,(ik \cdot R)\,\psi_k\,(r) \qquad\qquad (1.7)$$

Thus for a wave vector G satisfying:

$$G \cdot R = 2n\pi \qquad\qquad (1.8)$$

where, n is an integer, the eigenfunctions $\psi_k(r)$ and $\psi_{k+G}(r)$ are indistinguishable. The vector G is called a "reciprocal lattice vector" or "Umklapp vector." It may be recalled that a reciprocal lattice is dual to the usual crystal lattice in the real space, i.e., the basis vectors of the two lattices are related by the equations:

$$b_1 = 2\pi\,\frac{a_2 X a_3}{V} \qquad\qquad (1.9)$$

where, a_1, a_2, and a_3 are the basis vectors of the crystal lattice or Bravis lattice and b_1, b_2, and b_3 are the corresponding basis vectors of the reciprocal lattice; V is the volume of the unit cell in the crystal lattice.

This result bringing out the very definite functional form and properties for the eigenfunctions for a single electron moving through a periodic lattice is known as Bloch's Theorem—named after the celebrated physicist Felix Bloch who first showed this. The eigenfunctions $\psi_k(r)$ are called Bloch's functions. They represent the propagating solutions for the electron waves (De-Broglie waves corresponding to the electrons). These Bloch functions are also the 'electronic energy band states.' The corresponding eigenvalues ε_k are the 'electronic band energies.' They form band spectra of energy eigenvalues, in contrast to the discrete energy eigenvalues exhibited by the individual atoms. We will throw more light on this aspect later.

It is important at this stage, to get a feeling for the mathematical *viz.* analytical form of these band state wave functions and also of the *k*-dependence of the band energies. These can be achieved by choosing a simple model for the periodic potential V *(r)*. One such model is the 'Kronig-Penney model.' In this model, V *(r)* is approximated as Dirac-delta functions centered about the lattice sites. This leads to the following form for the Schrodinger equation in the simplest case of a one-dimensional lattice:

$$\left[-\nabla^2 + W \sum_i \delta\left(x - x_i\right) - k^2 \right] \psi_k\left(x\right) = 0 \qquad (1.10)$$

where, the periodic potential has been chosen as:

$$V(x) = \frac{h^2}{\left(2\pi\right)^2 2m} W \sum_i \delta\left(x - x_i\right) \qquad (1.11)$$

and k^2 being equal to $\dfrac{2m\left(2\pi\right)^2 \in_k}{h^2}$.

The consequences of this equation are striking! They show the existence of 'allowed energy values' and the 'forbidden energy values' within the *k* space. Furthermore, these sets of energy values form distinct band spectra of energy, *viz.* allowed band and forbidden band. In the language of wave mechanics, the electron eigenfunction corresponding to an allowed band is a propagating wave categorized as a Bloch wave; whereas the eigenfunction corresponding to a forbidden band is a damped wave. This result can easily be generalized to higher dimensions including three-dimension. In this case, the allowed *k* space is defined as the region with all the three components k_x, k_y, and k_z lying within $-\dfrac{\pi}{a}$ to $+\dfrac{\pi}{a}$, because of the existence of G. This minimum unit of *k* space region is known as the Brillouin Zone. The allowed energy bands may be fully occupied or partially occupied by electrons; or may be completely empty. We will see later that the partially occupied energy band with the highest energy is very important for the electronic transport properties of a material. It plays the most important role in deciding whether the material is metallic or insulating. This particular band is called the "conduction band." All the completely occupied (filled) allowed bands with energies below that of the conduction band are known as "valence bands."

The degree of occupancy of an energy band depends upon the number of electrons per each atom, belonging to the atomic orbital which forms that particular band. For example, take the case of solid Sodium. A Sodium atom has 1 lone electron in its outermost 3s orbital. Thus when a large number of Sodium atoms are brought very close to each other to form a solid Sodium, the discrete 3s orbitals of the atoms overlap and split the degeneracy of these atomic states to form a 3s band which becomes half-filled. The lower orbitals viz. 1s and 2s form the corresponding bands which are completely filled. Thus for solid Sodium, the 3s band forms the conduction band and the two lower bands become the valence bands. This half-filled (and half-empty) conduction band of Sodium has sufficient volume in the ε–k space for the electrons to get accelerated under the application of an external electric field. This leads to an excellent electrical conductivity for Sodium and its observed metallic properties. In general, it can be shown that metals always have an unfilled conduction band. Moreover, the metals with exactly half-filled conduction band are the best conductors. The metals formed from Group I elements of the Periodic Table belong to this class. The solids formed from Group II elements also behave like metals, even though according to the Kronig-Penney model they should not. The reason behind this is that the periodic potential V(r) in a real solid deviates from the idealized Kronig-Penney potential. As a result, real metal shows overlapping bands. This facilitates the creation of unfilled conduction bands for various solids made out of even some of the Group II and III elements and leads to metallic behaviors of these solids. The solids formed from other remaining elements are insulators, semiconductors, or semimetals.

One of the simplest approximations to the band Hamiltonian is that of the nearest neighbor "tight-binding approximation." In this approximation, the Hamiltonian, known as the tight-binding Hamiltonian is given by:

$$H = \sum_{i,\sigma} \epsilon_0\, c_{i,\sigma}^+ c_{i,\sigma} + \sum_{<i,j>,\sigma} t c_{i,\sigma}^+ c_{j,\sigma} \tag{1.12}$$

Then the ε(k) dispersion relation takes the following form for a three-dimensional semi-cubic lattice:

$$\epsilon(k) = \epsilon_0 - 2t(\cos k_x a + \cos k_y b + \cos k_z c) \tag{1.13}$$

where, t is the electron hopping matrix element involving nearest-neighbor lattice sites; a, b, and c are the lattice parameters along with the 3 crystallographic directions. The parameter ε_0 is the electron energy corresponding to an atomic site. The quantities t and ε_0 are determined by evaluating the matrix elements of the original 1st quantized Hamiltonian as used in Bloch's theorem (see Eqn. (1.1)), between the atomic orbital wave functions corresponding to the nearest neighbor sites and the same site respectively. The dispersion relation is given by Eqn. (1.9) shows that the process of hopping removes the degeneracy of the electronic energy spectrum of isolated atoms and causes the formation of an energy band with energies lying between $\varepsilon_0 - 6t$ to $\varepsilon_0 + 6t$ in a three-dimensional solid. The energy width of $12t$ between the bottoms of the band to the top of the band is known as the energy bandwidth.

In this most common version of the tight-binding model, the electron hopping is considered only within the nearest neighbor lattice sites. In the more general version, the hopping is considered even between the next nearest neighbor sites as well. This model works very well for Group II and III metals, semiconductors, and insulators.

The calculation of the dispersion relation or the band structure is computationally quite involved for a real material. To make the computational part simplified yet yielding accurate results, one takes recourse to the introduction of a quantity known as "pseudo-potential." The pseudo-potential, as determined by a proper choice of the electronic basis states, approximates the actual lattice periodic potential $V(r)$ for a real condensed matter system; however, it ensures that the band structure obtained is genuine and correct, particularly for the higher valence and the conduction bands. This method neglects the low-lying bands of the core electrons completely (Madelung, 1995).

From a very generalized band dispersion, as described earlier, it is possible to introduce the concept of "effective mass" or "band mass" of an electron. This represents the effective inertial mass of a band electron. This is defined as the 2nd derivative of the band energy with respect to the wave-vector, in the vicinity of a band edge, i.e., near the minimum or the maximum of the conduction band. More precisely, the effective mass tensor is given by:

$$m^*_{\alpha\beta} = \frac{h^2}{\left(2\pi\right)^2 g_{\alpha\beta}} \tag{1.14}$$

where,

$$g_{\alpha\beta} = \left[\frac{\delta^2 \epsilon_k}{\delta k_\alpha \delta k_\beta} \right]_{k_0} \qquad (1.15)$$

where, k_0 gives the position of the band edge in the allowed k-space and α and β are the Cartesian components.

From the above expression, we see that the effective mass becomes a k-dependent anisotropic function and a tensor quantity. For a simple tight-binding model corresponding to a simple cubic crystalline solid, however, the effective mass becomes a scalar and similar to that of a free particle. We may call this mass as $m_{quasi-free}(m_{qf})$. This is given by (from Eqn. (1.13)):

$$m_{quasi-free} = \frac{h^2}{(2\pi)^2 2ta^2} \qquad (1.16)$$

Thus, we see that m_{qf} is inversely proportional to the hopping parameter t. This is what is expected on the physical grounds; as larger, the hopping parameter is, the lighter the electron would appear to be and vice versa. This m_{qf} is generally larger than the real free-electron bare mass even for good metal. An interesting feature of m^* corresponding to any general dispersion is that near the upper extremity of a band, it becomes –ve. This leads to a well-known Bragg reflection occurring near the zone boundary.

The idea of the effective mass or band mass is also very useful in some of the semi-classical descriptions for the motion of a band carrier. A band electron is best represented by a "Bloch wave packet," constructed out of a suitable superposition of the extended Bloch wave functions around a particular wave vector. The quantum mechanical equation of motion of such a wave packet can be cast in the form of Newton's Second Law, by making use of the band mass and the group velocity – to be determined from the band dispersion by evaluating $\dfrac{2\pi d \in (k)}{h dk}$ – of the wave packet. This interesting result is known as "Bloch's acceleration theorem." This helps to understand the transport process involving the carriers in a band.

A very important function used extensively in band theory to represent electronic states, is the "Wannier function." This is obtained from Bloch's function by a mathematical transformation similar to Fourier transformation, viz.

$$a(R_n, r) = \frac{1}{N^{1/2}} \sum_k \exp(-i k \cdot R_n) \psi(k, r) \qquad (1.17)$$

where, R_n is the position vector of the nth atomic/ionic site in the lattice. The Wannier function $a(R_n, r)$ represents a wave function strongly peaked at the nth lattice site, in contrast to the Bloch function $\psi(k, r)$ which is completely delocalized in coordinate space. The Wannier functions also show oscillations and resemble the atomic wave functions to a very great extent; however, their added advantage is that they form an orthonormal complete set just like the Bloch functions. Thus for rigorous and very accurate calculations of the band structure, the various parameters occurring in the electronic Hamiltonians for the solid, like the tight-binding Hamiltonian, are in fact the matrix elements to be evaluated using the corresponding Wannier functions rather than the atomic orbital wave functions.

The various possible filling fractions of the bands also lead to an important concept *viz.* that of a "fictitious" but very useful particle called "hole." A hole in a band state k is visualized as a particle representing the absence of an electron in the k-state. Its effective mass and charge turn out to be exactly opposite to those of an electron. It can be shown rigorously that all the properties of a solid based on band theory, can be described in terms of electron or hole picture equivalently for both qualitative and quantitative understanding. Besides, in semiconductors, the electrical transport is governed by the concentration and the mobility of both electrons and holes from the conduction band and the valence band respectively. This proves the real and physical existence of the holes in band theory.

From the ε–k dispersion relation, whether from the tight-binding model or a general one, it is lead to the idea of a constant energy surface for the band electrons (and holes). This surface constructed in the k space is known as the Fermi Surface. The shapes of the Fermi surfaces vary widely from material to material. They may be closed surfaces or open surfaces. The exact shapes of the electron or hole Fermi surfaces of a crystalline solid depends on the detailed band structure of the material involving the magnitudes of the various material parameters occurring in the dispersion relation and also the valency of the parental element. For a genuine free-electron system, the Bloch function reduces to a plane wave with $u_k(r)$ becoming a constant and the Fermi surface is a sphere. For the monovalent alkali metals like *Li*, *Na*, *K*, etc., the Fermi Surfaces are approximately spherical, justifying the validity of nearly free electron approximation for these materials. For band electrons (and holes) in general, however, the topology of the Fermi surfaces can be very complicated and interesting. This is most clearly seen for some of the heavier alkali metals, polyvalent metals, and also the transition metals. The equation defining a Fermi surface can be written as:

$$\epsilon_k = Constant \tag{1.18}$$

The solution to the above equation, even for a simple tight-binding model in general, will have to be obtained numerically. The detailed investigation would lead to the various shapes and topologies of the Fermi surfaces, as discussed earlier. The Fermi surface corresponding to special characteristic energy called the Fermi energy is very important for metal, as this determines most of the electronic properties of the metal at least at low temperatures. The Fermi energy of a metallic system is defined as the highest energy level occupied in the conduction band at zero temperature. For a semiconductor and insulator, however, the Fermi energy falls well within the forbidden gap region. In fact, for an intrinsic semiconductor like pure Si and pure Ge, the Fermi level is situated exactly at the mid gap position. All these properties follow from the Fermi distribution function obeyed by the non-interacting band electrons and holes (Kittel, 1953).

A related quantity is often used in condensed matter physics to understand various phenomena and processes of electronic origin is the "electronic density of states (EDOS)." It is defined as the number of band states within an energy interval of ε to $\varepsilon + d\varepsilon$. More precisely, mathematically it is defined as:

$$N(\epsilon) = \sum_k \delta\left(\epsilon - \epsilon_k\right) \tag{1.19}$$

In particular, the density of states (DOS) evaluated at the Fermi energy of a metal, characterized by $N(0)$, is a very important parameter occurring in the expressions for various electronic properties of a solid, as we will see later. Within a free electron-like approximation (corresponding to jellium model which is a simplification of the lattice structure to a continuous jelly and is a fairly good approximation for the alkali metals), for a three-dimensional system $N(0)$ can be shown to be proportional to m^*. In general for a 3d lattice under this approximation, $N(\varepsilon)$ goes as $\epsilon^{\frac{1}{2}}$.

For a 1-dimensional lattice, the jellium model approximation gives $N(\varepsilon)$

proportional to $\epsilon^{-\frac{1}{2}}$ and for a two-dimensional system it is a constant. Again in general for an electronic band, $N(\varepsilon)$ can be quite a complicated function of s depending upon the band dispersion. Moreover, in such a case, $N(\varepsilon)$ will have to be evaluated numerically. For a nearest-neighbor tight-binding model, however, it is possible to obtain an exact analytical form for $N(\varepsilon)$ in 1d and 2d. They deviate quite strongly from those obtained with the free-electron model. In particular, for a 2d square lattice with symmetric nearest-neighbor hopping along with the 'a' and 'b' directions, the DOS takes the form of an elliptic integral and exhibits a singularity at some special value of ε. Inclusion of the hopping contributions for higher neighbors leads to more complicated expressions. This has tremendous consequences for various condensed matter phenomena particularly at low temperatures (Roy Chowdhury and Chaudhury, 2016, 2018).

So far, we have confined our discussions only to a perfectly crystalline solid. In reality, there are always random impurities and defects present in a lattice. This causes the translational invariance of the electron-lattice potential $V(r)$ to be broken inside the solid. Thus, this amounts to a violation of Bloch's theorem. Moreover, a finite-sized solid also suffers from the discontinuity of the potential $V(r)$ at the boundary, leading to the emergence of new eigenstates different from the Bloch states. All these spatial disorders related effects give rise to the existence of a number of stable electronic states even within the forbidden region or the bandgap region of the usual band theory corresponding to the ideal lattice. This aspect will be taken up in greater detail in Chapter 4 in the section on "Surface States."

The understanding of the effect of Coulomb interactions between electrons is extremely important in the conduction band to explain various properties of solids observed experimentally. This is however quite non-trivial in terms of theoretical treatment. Before we come to this, let us have a look at the interacting electron gas problem. It is well known from the perturbative treatment of high-density electron gas that the Coulomb interaction amongst electrons essentially leads to Fermi-liquid theory where the basic properties of free electron gas remain intact with electron mass scaled from m to m^*. With the band electrons, this analysis becomes much more complex.

We introduce and devote some clarifying discussions on some of the concepts like Hartree-Fock, correlation energy involving perturbation treatment in high density and metallic density electron gas, Wigner solid

(WS) formation in low-density electron gas and its melting, etc., taking care of many-body effects in electron gas and in band structure calculations; Fermi liquid theory, Tomonaga-Luttinger liquid (TLL) theory; DFT with Kohn-Sham approach for exchange-correlation contributions; Karr-Perinnello scheme for the electron-ion system; topological insulators (with bulk insulating and surface metallic), graphene, carbon nanotubes, tunneling in the topological insulator-superconductor junction (involving a junction of the topological insulator and a superconductor in two-dimensions with the emergence of so-called "Majorana fermions" in Chapter 4 in some details (Pines, 1999; Mahan, 1981; Atland and Simons, 2006).

1.3 EXPERIMENTAL VERIFICATION OF BAND THEORY AND DETERMINATION OF FERMI SURFACE

The experimental determination of the shapes of the Fermi surfaces is most commonly done by making use of the De Haas-van Alphen effect, arising from the diamagnetic response of the band carriers (electrons or holes). This method involves tracking the motion of band electrons (or holes) in a static uniform magnetic field. As an electron (or hole) experiences the Lorentz force $qvXB$, its motion always takes place on a surface of constant energy. Thus by this method, one can scan the Fermi surface easily. The detailed discussion regarding the De Haas-van Alphen effect will be carried out in the section on "magnetism." For the time being, it is sufficient to say that in this effect, the diamagnetic susceptibility of the band electrons (or holes) in the high magnetic field and at low temperature, shows an oscillation with reciprocal of the magnetic field. The origin of such oscillation lies in the progressive shifting of the Fermi energy level as well as the alternate populating and de-populating of the discrete energy levels (known as Landau levels and generated by the external magnetic field) by the electrons/holes, as the magnetic field increases. The period of this oscillation depends upon the shape and other parameters of the Fermi surface.

Another phenomenon closely related to De Haas-van Alphen effect is that of "cyclotron resonance." The carriers in the band, when subjected simultaneously to a static uniform magnetic field and also to an electromagnetic (e.m.) field, exhibits a resonance absorption of the electromagnetic energy at a particular frequency of the e.m. wave. This frequency is known as the cyclotron frequency. The cyclotron frequency turns out to be inversely proportional to the effective mass (m^*) of the band carrier. Thus from the measurement of cyclotron frequency, one can determine m^*. This in turn provides valuable information about the dispersion relation of the energy bands. The detailed discussions regarding this will be carried out in Chapter 3, dealing with the interaction between the electron and electromagnetic waves.

Besides the above two experimental techniques, "photoemission spectroscopy" along with a more powerful angle-resolved photo-emission spectroscopy and spin-resolved photo-emission spectroscopy, are also used to determine the detailed shape and nature of the Fermi surface with spin correlations present, in various materials. It can also in general provide information regarding the energy dispersion of various bands. The absorption spectroscopies of different kinds also prove the experimental existence of the electronic energy bands with the band gaps (Kittel, 1953; Madelung, 1995).

The effects of a uniform static magnetic field on the transport properties of band carriers in a metal or semiconductor can be quite diverse! Two notable phenomena related to this are: (i) magneto-resistance and (ii) quantum Hall effect (QHE). The magneto-resistance in a paramagnetic metal is primarily due to the loss of the conduction electrons participating in the drift current, due to the Lorentz force exerted by the magnetic field on these electrons. Thus in an experiment involving a combination of a dc electric field and a magnetic field along the perpendicular direction, the current along the direction of the electric field decreases with the increase in the strength of the magnetic field. This tantamounts to an electrical resistance induced by the applied magnetic field and are known as the magneto-resistance. Moreover, the accumulation of the deflected electrons leads to the formation of a potential difference, known as the Hall voltage, along a direction perpendicular to both the direction of the electric field and that of the magnetic field.

The other related effects are Hall effect, both classical and quantum. In the classical Hall effect seen in a three-dimensional conductor (metal or semiconductor), the sign of Hall voltage developed is an indicator of the nature of the charge carriers *viz.* electron or hole. This is due to the fact that the Lorentz force is the same for both these two types of carriers when contributing to the current and as a result, the Hall voltage developed between the two electrodes placed will be of opposite sign for these two types of carriers. While in a metal, the charge carrier is always an electron, in a semiconductor it can be either an electron or a hole. Moreover, the Hall coefficient (derived from the ratio of Hall voltage to Hall current) gives the carrier concentration also.

In QHE, seen in a conductor of highly two-dimensional geometry, the transverse Hall conductance (defined as the ratio of usual transport current to Hall voltage) plotted as a function of the magnetic field shows a series of plateau. Because of this plateau structure, Hall's conductance is quantized in units (integer or fraction) of the fundamental conductance $\dfrac{\left(e^2\right)2\pi}{h}$.

This is a purely quantum-mechanical effect originating from the existence of discrete electronic energy levels *viz.* Landau levels and also from the presence of disorder and sometimes even many body effects. The experimental discovery of integer QHE (integer multiple of the fundamental conductance) was made by Alexander Von Klitzing at IBM (Zurich) and was awarded Nobel Prize in 1985 for this. The experimental system was a highly pure two-dimensional electron gas confined in a hetero-junction involving *GaAs –GaAlAs* with parts per million (ppm) level of impurity concentration. Theoretical understanding of these phenomena has been quite successful (Pruisken, 1987; Laughlin, 1981; Jain, 1989).

In recent times, there has also been a lot of interest in a phenomenon known as the "spin Hall effect." This is related to the generation of spin current in the presence of an applied electric field (or electric current) in some of the materials which exhibit strong spin-orbit coupling of the form of Dzyaloshinskii-Moriya (DM) interaction (Dyakonov and Perel, 1971; Sinova et al., 2015) (Figure 1.1).

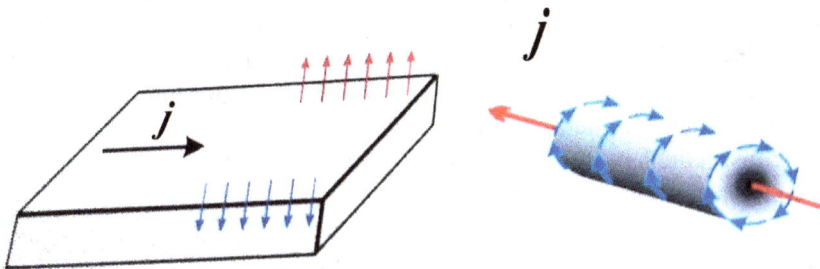

FIGURE 1.1 The spin Hall effect. An electrical current induces spin accumulation at the lateral boundaries of a sample.

Source: Reprinted with permission from Dyakonov (2009, 2012).

1.4   VARIOUS NOVEL PHENOMENA ASSOCIATED WITH FERMI SURFACE

There are several other important phenomena and effects associated with the Fermi surface. Some of these are: Pauli paramagnetism, Stoner paramagnetism, spins density wave, charge density wave (CDW), and superconductivity. Since all these phenomena occur or originate in metals, here the Fermi surface with energy equal to the Fermi energy is only relevant. The phenomenon of Pauli paramagnetism is a temperature-independent paramagnetic susceptibility exhibited by metals at low temperature. The magnitude of the spin susceptibility is proportional to the carrier density of states at the Fermi energy ($N(0)$). Stoner paramagnetism is an enhanced form of Pauli paramagnetism, the enhancement arising from the presence of the many-body, i.e., Coulomb interaction between the electrons. An elaborate analysis of both Pauli and Stoner paramagnetism will be presented in Chapter 2 dealing exclusively with magnetism. Many of these phenomena can be discussed qualitatively and sometimes semi-quantitatively even within the jellium model approximation, without taking recourse to the detailed band structure.

The SDW occurs when a metal because of "many-body effects" arising predominantly from Coulomb interactions amongst the band electrons, develops an energy gap at the Fermi surface. The metal becomes a semiconductor and instead of Pauli or Stoner paramagnetism, one observes a magnetically ordered state with an oscillatory spatial distribution of magnetic moments. Sometimes the SDW may occur also because of competition between different possible magnetic orderings involving the band electrons/carriers, due to the complexities of interactions at the Fermi surface. In an analogous way, the CDW also develops in a metal undergoing a transition to a semiconductor, due to the energy gap developing at the Fermi surface from electron-lattice interaction. It should be remarked however that the occurrences of both the SDW and the CDW need the Fermi surface to possess certain symmetry known as "nesting." This is found to be satisfied generally in the low-dimensional metals. The detailed analysis of these two phenomena can be found in Chapter 2 and Chapter 3, respectively.

In comparison with the phenomena described above, the phenomenon of superconductivity is a very different one. Although originating from instability at the metallic Fermi surface leading to the formation of an energy gap, the DC electrical resistivity vanishes completely when the metal is cooled below a certain temperature. It should be emphasized here that the nature of the energy gap developed here is drastically different from that seen in a semiconductor. The energy gap in a superconductor arises from an effective electron-electron (carrier-carrier) attraction brought about generally by the electron-lattice interaction in a metal. As a result, two electrons near the Fermi surface pair up to form an apparent bound state, and the normal metal goes over to a superconductor. This pair of electrons known as a Cooper pair moves as a whole coherently in the superconducting phase. Thus, this transport process avoids Ohmic dissipation and the system exhibits an infinite d c electrical conductivity. The temperature, at which the transition from a normal metal to a superconductor takes place, depends upon the range and the strength of the attractive interaction between the electrons as well as on $N(0)$. The detailed analysis of the origin and the properties of superconductors will be dealt with in Chapters 3 and 5.

Very recently there have been a lot of experimental and theoretical activities related to a certain new type of ferromagnetic materials known as "Heusler alloys, discovered in 1903 by Heusler (Webster, 1969). In many of these materials, amongst the band carriers, the majority spin sub-band has a finite EDOS at the Fermi energy and hence exhibits a metallic character; whereas the minority spin sub-band develops an energy gap in the electronic excitation spectrum at the Fermi level, displaying an insulator-like behavior. Therefore, they are also known as "half-metals," like for example $NiMnSb$. As a consequence, both the magnetic and transport properties of half-metals are very unusual and fascinating. The most notable aspect of the Heusler alloys is that these alloys mostly ternary, are ferromagnetic whereas the primary components of these alloys are non-magnetic, like for example Cu_2MnAl. These would be taken up again in Chapter 2. The newly emerging applied field of "spintronics" makes quite heavy use of such half-metallic behavior. An allied phenomenon is known as tunneling magneto-resistance (TMR) which shows the existence of very different magnitudes of electrical resistance for inter-layer charge transport corresponding to the parallel and anti-parallel spin relative orientations for the carriers in the two layers is an active field of research for the experimentalists and technologists.

Before concluding this chapter, it may be worthwhile to draw attention to two more closely related issues $viz.$ that of (i) heavy fermion metals and superconductors and (ii) strongly correlated electronic systems, which are exotic systems.

In the heavy-fermion materials, the effective mass of the conduction electrons is found to be gigantic. The ratio of the observed m^* to the usual band mass sometimes exceeds even 1000. As a result, the Fermi energy is very low and the EDOS at the Fermi surface is very high compared to the usual metals. These systems also show a unique crossover from localized magnetism (arising from the localized atomic-like electrons) to itinerant magnetism (due to the truly de-localized band electrons) as temperature goes down and at very low temperature; some of them even become super-conductors. Moreover, some amongst the non-superconducting metals also exhibit enhanced paramagnetism of Stoner type. In these systems, which are Lanthanides and Actinides and are made up of rare earth and transition metal elements, the detailed analysis shows a very high degree of hybridization between the s or d energy bands from the transition metals and the f atomic state s of the rare earth elements. The many body effects *viz.* the direct Coulomb interaction as well as that mediated through the lattice, between these electrons, also play quite important roles in the properties of these systems. In particular, the mechanism and the nature of superconductivity in these materials are still not understood completely. We will have a more elaborate discussion on these aspects in Chapter 2.

The strongly correlated electronic systems are those systems where the conventional band theory breaks down. The transition metal oxides and the cuprates exhibiting high-temperature superconductivity belong to this class of systems. In these materials, the one-electron treatment based band theory fails. This is because the strong Coulomb interaction matrix element involving two electrons with opposite spins belonging to the same lattice site, known as Mott-Hubbard repulsion, is much larger than the single electron hopping matrix element connecting two nearest-neighbor sites. This makes the electronic conduction energetically unfavorable. As a consequence, a transition metal oxide like NiO having 1 electron per site (equivalent to a half-filled band in normal band theory), turns out to be an insulator rather than a metal. Again the underdoped rare earth-based cuprates like $La_{2-x}Sr_xCuO_4$ for x below 0.1, behave as very unconventional strongly correlated metals (conductors), exhibiting properties not expected from standard band theory with minor corrections from weak inter-electron interactions. These anomalous conducting systems are derived from the parental systems like CuO which are insulators (for reasons similar to those corresponding to NiO and probably also due to some other many-body processes), by the creation of vacancies and thereby facilitating some amount of electronic conduction. In these unconventional conductors, however, even the Fermi surfaces are not well defined sometimes. As a result, the known phenomenological models describing the conventional metals cannot be applied here. Consequently, the theory of superconductivity for these materials is very complex. This is analogs to Wigner Crystal and its melting, as discussed in the context of the jellium model. We will take up some of these issues in Chapters 4 and 5.

Lastly, even the characterization of the materials on the basis of electrical polarization properties need not be dependent on the behavior of the ionic components only. The processes which are responsible for the formation of CDWs and SDWs, may also lead to ferro or anti-ferroelectricity of electronic origin (Ginzburg and Kirzhnits, 1982). The detailed theory often involves a scenario of two electronic energy bands with different levels of filling corresponding to a semi-conductor or semi-metal. In this context, it is also interesting to comment that quite recently another new class of materials known as "bi-ferroics" have been prepared in laboratory (Manfred and Lei). They are mostly insulating however, some of them can be of metallic character as well. They exhibit coexistence or a crossover between ferromagnetic and ferroelectric phases. More generalized systems known as "multi-ferroics" have also been synthesized (Khomskii, 2004, 2006; Cheong and Mostovoy, 2007; Ray and Waghmare, 2008). They display very rich interplay involving magnetism, ferro (or antiferro)-electricity, and ferro-elasticity via Piezo-electric effect and magnetostriction. These features can occur on both microscopic as well as macroscopic scales. Again, the spin-orbit coupling in the form of DM interaction or Rashba coupling plays a very important role in the origin of the complex combined spin and charge responses of these classes of systems. Some of these materials with layered structures and with very high spin polarization of the conduction electrons, like Manganite-Titanate composites, can also find a lot of promising technological applications in various novel devices based on "spintronics" and "spin valve" (Dorr and Thiele, 2006). The field of spintronics deals with spin transport electronics as opposed to charge transport in various solid-state devices (Yuassa et al., 2004; Parkin et al., 2004). In particular, this involves exotic phenomena such as TMR and GMR (Giant magneto-resistance) which are found to occur in sandwich structures comprising of alternate ferromagnetic and non-magnetic metal layers in experiments. The above effects have turned out to be very useful for applications in the memory devices. Besides, another general class of 'novel' or 'smart' materials called "shape memory alloys" (ordinary or magnetic) have gained considerable interest amongst the condensed matter physicists and technologists. These systems display the memory effects related to the physical shape or magnetization with the application and withdrawal of mechanical stress or magnetic field under the thermal cooling and heating cycle. The understanding of these phenomena requires a detailed analysis of lattice structure with defects or distortions, the magnetic domains, and coupling between the two as well. An example of such material is Ni_2MnGa (Wayman, 2013).

Finally, the recent experimental and technological advances in the areas of nano and mesoscopic sciences with the development and investigation of single-walled carbon nanotube (SWCNT), graphene, and quantum dots have contributed a lot in enriching the fields of materials sciences as well as condensed matter physics, particularly in the areas of theory and computation of electronic structures. The graphene sheet is a single layer of graphite, which possesses a narrow band of π-orbital states. This can be modeled by a tight-binding Hamiltonian (TBH) of electronic band theory as a first approximation. A striking feature of the resulting electronic structure is that the half-filled system is characterized by a point like the Fermi surface. Furthermore, when slightly doped to move away from the half-filling, the system exhibits two Dirac-like electron spectra with a linear ε–k dispersion. The major practical utility of graphene lies in the fact that it is a very light, i.e., low-density material although very strong (Greim and Kim, 2008; Atland and Simons, 2006). A SWCNT is a one-dimensional structure involving a graphene sheet rolled into a cylinder. Again, electronic structure calculations based on TBH show the presence of two nodal points in SWCNT and the electronic spectrum shows approximately linear dispersion in the vicinity of one of these nodal points (Atland and Simons, 2006). Technological advancements have enabled the manufacturing of mesoscopic devices (of sizes of the order of 10^{-4} centimeter) of metallic or semi-conducting character. A quantum dot is one such device. Charging of a quantum dot is a very interesting process which exhibits vanishing of EDOS at the Fermi surface due to the Coulomb Blockade effect, arising from inter-electron interaction. This has a tremendous application towards the tunneling characteristics for a 2D electron gas formed at the interface between $GaAs$ and $AlGaAs$ layers, behaving as a quantum dot. Again, the realistic systems belonging to the above class all contain spatial disorder weak or strong, and further are non-self-averaging due to the mesoscopic sizes. These aspects are manifested in various novel phenomena like Anderson's localization, different types of anomalous transport behavior involving zero-bias anomalies, both cooperon and diffusion contributions to charge current, etc., (Efetov et al., 1982; Kamenev, 2004; Edwards and Anderson, 1975; Altshuler et al., 1982). In many of these systems, the combined effect of disorder and many-body interactions produce unexpected outcomes and both perturbative, as well as non-perturbative treatments, become essential for the clear understanding (Altshuler et al., 1982). A quantum mechanical equation for charge transport in mesoscopic systems was originally developed by Landauer and Buttiker, although it had several limitations like the neglect of dissipation at the junctions or leads (Leonhard, 2010). A statistical mechanical treatment leading to Kubo-like formula for linear response, in this case, was derived later with the help of Keldysh Green functions and with the inclusion of dissipation (Das, 2010). Some of these issues will be taken up briefly in Chapter 4 again.

All these are glaring examples of the confluence of material science, technology, and condensed matter physics with involvement and application of both band theory, many-body physics, and statistical mechanics.

KEYWORDS

- density of states
- electronic density of states
- giant magneto-resistance
- quantum hall effect
- single-walled carbon nanotube
- tunneling magneto-resistance

REFERENCES

Altshuler, B. L., Aronov, A. G., Khmelnitskii, D. E., & Larkin, A. I., (1982). Chapter 3. In: I. Lifshits, M., (ed.), *Quantum Theory of Solids*. MIR Publishers, Moscow.

Ashcroft, N. W., & Mermin, N. D., (2017). *Solid State Physics*. Brooks/Cengage Learning India.

Atland, A., & Simons, B., (2006). *Condensed Matter Field Theory*. Cambridge University Press, Cambridge.

Capra, F., (1982). *The Tao of Physics*. Flamingo (Harper Collins Publishers), London, Chapter 11.

Casper, F., Graf, T., Chadov, S., Balke, B., & Felser C., (2012). *Semiconductor Science and Technology, 27*, 6.

Chaudhury, R., & Jha, S. S., (1984). *Pramana Journal of Physics 22*(5), 431.

Chaudhury, R. (2009). *Cond-mat arXiv.*

Cheong, S. W., & Mostovoy, M., (2007). *Nature Materials, 6*, 13.

Das, M. P., & Green, F., (2003). *arXiv: Cond-mat/0304573.*

Das, M. P., (2010). *IOP Science*. iop.org/article/10.1088/2043-6262/1/4/043001/.

Dorr, K., & Thiele, C., (2006). https://doi.org/10.1002/pssb.200562441.

Dyakonov, M. I., &Perel, V. I., (1971). *Sov. Phys. JETP Lett., 13*, 467; *Phys. Lett. A., 35*(6), 459 (1971).

Dyakonov, M. I., (2009). *arXiv: 1210.3200 [cond-mat.mes-hall] (2012)* (Vols. 2–5, p. 656). Google Scholar.

Economou, E. N., (1990). *Green's Functions in Quantum Physics*. Springer Verlag, Heidelberg.

Edwards, S. F., & Anderson, P. W., (1975). *J. Phys. F5.*

Efetov K. B., et al., (1982). *Sov. Phys. JETP, 55*, 514.

Geim, A., K., & Kim, P., (2008). *Scientific American, 298*(4), 90.

Ginzburg, V. L., & Kirzhnits, D. A., (1982). *High Temperature Superconductivity*. New York Consultants Bureau, New York.

Jain, J. K., (1989). *Phys. Rev. Lett., 63*, 199.

Kamenev, (2004). *Cond-Mat/0412296.*

Khomskii, D. I., (2006). *Journal of Magnetism and Magnetic Materials-2004*. https://arXiv.org/pdf/Condmat/0601696 (accessed on 7 July 2020).

Kittel, C., (1953). *Introduction to Solid State Physics*. Wiley, New York.

Laughlin, R. B., (1981). *Phys. Rev. B, 23*, 5632.

Leonhard, N., (2010). *Conductance Quantization and Landauer Formula, SS.*

Madelung, O., (1995). *Introduction to Solid State Theory*. Springer, Heidelberg.

Mahan, G. D., (1981). *Quantum Theory of Many-Particle Systems*. Plenum Press.

Manfred, W., & Lei, Z. *Cobaltferrite-Bariumtitanate Sol-Gel Biferroics*. Digital Repository at University of Maryland.

Pines, D., (1999). *Elementary Excitations in Solids*. Advanced books classics, Perseus Books Publishing, Massachusetts.

Pruisken, A. M. M., (1987). Field theory, scaling, and the localization problem. In: Prange, R. E., & Girvin, S. M., (eds.), *The Quantum Hall Effect*. Springer-Verlag.

Ray, N., & Waghmare, U. V., (2008). *Phys. Rev. B., 77*, 134112.

Sinova, J., Valenzuela, S. O., Wunderlich, J., Back, C. H., & Jungwirth, T., (2015). *Rev. Mod. Phys., 87*, 1213.

Wayman, C. M., (2013). *MRS Bulletin*. doi.org/10.1557/S0883769400037350.

Webster, P. J., (1969). *Contemporary Physics, 10*(6), 559.

Yamada, K., (2004). *Electron Correlation in Metals*. Cambridge University Press, Cambridge.

CHAPTER 2

Magnetism

2.1 INTRODUCTION

The phenomenon of "magnetism" is one of the oldest phenomena discovered in the field of condensed matter physics. Besides the discovery of magnetic stone (loadstone) by Magnes (2000 BC) and Gilbert (17th Century), it was also known since long back that the earth itself also acts as a magnet. The discovery of magnetic materials, more precisely ferromagnetic materials such as *Fe, Ni, Co*, etc., has contributed a lot to the progress of material science and technology. Besides the application aspects, the study of the phenomenon of magnetism has also enriched the basic sciences. To start with, the so-called "microscopic theory" for paramagnetism was attempted by the classical electromagnetic theory involving the magnetic dipoles representing the moments. The randomly oriented dipoles at a high temperature tend to get oriented by the application of an external magnetic field and thereby produce an induced magnetization. Langevin was the proponent of this idea. However, very soon this theory ran into problems. Later with the development of quantum ideas, there was an attempt to introduce "spins" to represent the moments by Brillouin. This was fairly successful. The theoretical prediction of the functional dependence of the induced magnetization with the applied magnetic field and temperature was found to be in quite good agreement with the experimental observations for various paramagnets.

After the successful treatment of the ordinary paramagnets, the efforts began for the understanding of the transition from a paramagnet to a ferro-magnet, seen in a large class of materials. It was realized very soon that the interaction between the magnetic moments plays a crucial role in this case. Thus, initially, the classical electromagnetic interaction between the magnetic dipoles was included in the model. However, the calculated transi-tion temperature turned out to be very small, *viz.* of the order of 4 degrees Kelvin. This is below the experimental values seen in most of the well-known systems. Furthermore, this model did not even ensure a parallel alignment of the moments always. Therefore, it was concluded that this classical dipole based model is not adequate to account for the ferromagnetic transition. Later it was shown mathematically by Van-Leeween from the considerations of equilibrium statistical mechanics that no form of magnetism, including diamagnetism, can be accounted for in a macroscopic system, by a truly microscopic theory within the purview of classical physics (White, 2007).

Werner Heisenberg with Paul M. Dirac, two of the founders of Quantum Mechanics, independently proposed in 1926 an alternative and purely quantum model for ferromagnetism, based on the electrostatic interaction between the electrons belonging to the atoms in a solid and the symmetry of the wave-function corresponding to these electronic systems. This model, known as the Heisenberg model, was somewhat along the lines of the theory for the formation of H_2 molecule by the Heitler-London approach. The Heisenberg model has been used to explain the origin of ferromagnetism in insulators. Later when the phenomenon of antiferromagnetism was discovered, the true quantum nature of magnetism was manifested more clearly. Subsequently, different quantum models were suggested corresponding to various types of magnetism observed in different materials. These models have been more or less successful both from the theorists' as well as experimentalists' viewpoints. Furthermore, these models are derived from very basic condensed matter physics Hamiltonians.

The plan of this chapter is as follows: In the first section, to start with we state and prove Van-Leewen's theorem. Then we discuss diamagnetism. In the second section, we present a theory for magnetism in insulators and semiconductors. We cover different types of magnetism *viz.* paramagnetism, ferromagnetism, antiferromagnetism, ferrimagnetism, etc. Then in the next section, we turn to itinerant magnetism, i.e., the magnetism of metals and metallic alloys. Here again, we attempt different types of observed *viz.* paramagnetism, ferromagnetism, antiferromagnetism, spin density wave (SDW), etc. In the fourth section, we present discussions for the magnetism of exotic types and in novel materials. This includes spin glass, magnetic superconductor, heavy fermion superconductor, doped cuprates, manganites, etc., in some of which one sees a cross over between localized magnetism and itinerant magnetism and also the interplay between magnetism and superconductivity. Moreover, in many of the above systems, both spin and orbital degrees of freedom contribute to magnetism in a very non-trivial way. We also touch upon the exciting problem of magnetism in the low dimensional systems, which are of paramount importance in both basic and applied sciences. In the final section, i.e., the fifth section we briefly discuss various experimental techniques used in the studies of magnetism. This covers neutron diffraction, inelastic neutron scattering, nuclear magnetic resonance (NMR), electron spin resonance (ESR), etc.

2.2   ORIGIN OF THE THEORY FOR MAGNETISM AND ANALYSIS OF DIAMAGNETISM

We start out with the theorem showing the impossibility of the occurrence of any form of magnetism in the classical theory. As stated before, this is Van-Leewen's theorem. The proof goes as follows: the truly microscopic theory involving elementary particles with electrical charges similar to those of electrons, is considered classically. The partition function for a thermodynamic system of such free particles is given by:

$$z = \int \pi_i \, dp_i dq_i \, \exp\left(-\frac{\beta}{2m} \sum_i p_i^2\right)$$

(2.1)

where, p_i and q_i are the momenta and the position vectors of these particles. In the presence of an external vector potential A, however, the momenta get transformed to $\left(P_i - \frac{eA}{c}\right)$. Let us denote these transformed momenta by p_i' and the partition function in the presence of this external potential as Z'. Then making use of Eqn. (2.1), one can easily see that Z' is identical to Z. Therefore, in the classical theory, the free energy of a system remains invariant when an external electromagnetic field is applied. Again, magnetization induced is proportional to the derivative of the free energy with respect to the external magnetic field. Thus, our simple calculation presented above shows that no magnetization is induced in a classical system even in the presence of an external magnetic field, in thermodynamic equilibrium.

The above analysis clearly implies that magnetism of materials, as observed, will have to be essentially a quantum phenomenon. Therefore, we will have to deal explicitly with the quantum mechanical Hamiltonians for describing all types of magnetism. To clarify this point, for example, even though the phenomenon of diamagnetism is a universal one and is based on the property of classical electromagnetic induction, the finite diamagnetic response is only obtainable in a fully quantum mechanical treatment of the material system which is coupled to the electromagnetic field. For the other forms of magnetism, however, the microscopic origin itself is quantum mechanical besides the necessity of using quantum variables in the Hamiltonian.

The phenomenon of diamagnetism or more precisely, orbital diamagnetism is a universal one, as stated before. Every system in this universe exhibits diamagnetic response, i.e., negative magnetic susceptibility. Its magnitude, i.e., the absolute magnitude of the diamagnetic susceptibility, however, is generally quite small and is very often masked by the higher magnitude of the spin response (positive magnetic susceptibility) from the other forms of magnetism like paramagnetism. The only exceptions are superconductors and some of the band insulators. These above-mentioned systems behave as "super-diamagnets," where the diamagnetic suscepti-

bility attains the highest value possible $viz. \dfrac{-1}{4\pi}$. Moreover, these systems have vanishingly small spin susceptibilities. As a result, the overall magnetic response of these systems is very prominently diamagnetic.

The origin of orbital diamagnetism is electromagnetic induction, as has been pointed out before. The applied magnetic field is opposed by the induced orbital screening current, governed by Faraday's Law or Lenz's Law. The simplest quantum mechanical model incorporating this effect was suggested by Landau among many others. The model essentially describes a collection of non-interacting electrons in the presence of an external electromagnetic field. It should be emphasized that the system of electrons is treated quantum mechanically but the electromagnetic (e.m.) field is regarded as classical. The Hamiltonian of such a system is given by, as used earlier in Eqn. (2.1):

$$H = \frac{1}{2m} \sum_i \left(\mathbf{p}_i - \frac{eA}{c} \right)^2$$

(2.2)

Here, the variable p_i is an operator denoted by $-i\dfrac{h}{2\pi}\nabla$. A is the e.m. vector potential experienced by the electrons. Simplifying the expression above, one can show that the term proportional to A^2 leads to the orbital diamagnetism. It should also be pointed out that the term proportional to operator p_i corresponds to the orbital paramagnetism and this issue will be taken up later in the next section. For electrons in an atom, the diamagnetic susceptibility calculated using the term proportional to A^2, can be shown to be proportional to the square of the average radius of the electronic orbital. The absolute magnitude of this susceptibility per unit volume is found to be extremely tiny *viz.* of the order of 10^{-6} in CGS units. Again, this diamagnetism generating a term occurring in Eqn. (2.2), can be recast in a very standard form as given below:

$$H_{coupl} = \frac{-1}{c} \int j(r) \cdot A(r,t) dr \tag{2.3}$$

where, the left-hand side represents the e.m. coupling between the electronic system and the external vector potential field and $j(r)$ is the orbital current density operator (for simplicity we have dropped the hat symbol on the operator). This form is the most common and widely used form of the perturbation term for calculating the electromagnetic response in a condensed matter system. This term can then be treated in linear response theory to extract the diamagnetic response function or the diamagnetic susceptibility for various condensed matter systems. For this, only the diamagnetic contribution to the orbital current density operator will have to be used.

The operator corresponding to this is given as:

$$J_{dia}(r) = -\sum_{i=1}^{N} \frac{e^2}{mc} A(r) \int \pi_i \, dr_i \, \psi^+ \psi \delta (r_i - r) \tag{2.4}$$

where, $\psi(r_1, r_N)$ is the field operator corresponding to the system. The detailed treatment brings out the role of the excitation spectrum of the condensed matter system. It can be shown explicitly that for systems possessing an energy gap between the ground state and the lowest excited state, like superconductors and some of the insulators/semiconductors, the absolute magnitude of the diamagnetic susceptibility per unit volume,

reaches the maximum possible value *viz.* $\frac{-1}{4\pi}$ in the CGS units, as stated

before. Moreover, the orbital and the spin paramagnetic susceptibility can be shown to be vanishingly small for them, as we will discuss later. Thus, they are the giant or super-diamagnets. The normal metallic systems, possessing well-defined Fermi surfaces and having gapless excitations, are however moderate diamagnets. Moreover, they also have finite spin susceptibility, as will be shown in later sections.

In the previous chapter, we have discussed briefly de-Hass-van Alphen effect and quantum hall effect (QHE). Both of these phenomena arise from the diamagnetic response. In addition, the spatial dimensionality of the system also plays an important role.

2.3 MAGNETISM IN INSULATORS

The magnetism in insulators owes its origin to the magnetism in the constituent atoms. A magnetic solid must have atoms or ions with the net magnetic moment, i.e., with an unfilled electronic shell having a net non-zero m_l (azimuthal quantum number) or m_s (spin quantum number). In other words, the solid must have paramagnetic atoms or ions. The existence or non-existence of the net moment on the atoms is essentially governed by the celebrated Hund's Rule of atomic physics. Of course, the situation is much more complicated in a solid compared to that in an isolated atom, as the electrons belonging to an atom experience the electrostatic fields from the nuclei and the electrons from other atoms. To be more precise, even the electrons which are essentially localized within the atoms in a solid, experience the so-called "crystal field." These changes the electronic energy diagram from that expected in an individual atom. Therefore, Hund's Rule will have to be applied on this modified energy level diagram to ascertain whether the condensed matter system will have magnetic atoms or not. The next question is whether the magnetic atoms are interacting among themselves or they are non-interacting. If they are non-interacting, the system exhibits paramagnetism, i.e., the absence of any spontaneous magnetization in the system. If they are interacting, however, then they may order magnetically below a certain temperature. The ordering can be of various patterns like ferromagnetic, antiferromagnetic, or little uncommon like ferrimagnetic, spin glass, etc. Nevertheless, all of these different phases show the appearance of an average non-zero moment per atomic site in the absence of any external magnetic field. The type of the magnetic order occurring will depend upon the nature of the interaction amongst the magnetic atoms. Above their respective ordering temperatures, however, all of these magnetic systems behave as paramagnets.

We now take up the topics related to each of the types of magnetic ordering listed above. Moreover, we also discuss various kinds of paramagnetism observed in materials *viz.* spin paramagnetism and orbital paramagnetism. Furthermore, based on the temperature dependence the spin paramagnetism can be further subdivided into three types *viz.* (i) Curie type, (ii) Curie-Weiss type, and (iii) Van-Vleck type for the insulators. The first and third type of paramagnetism belong to the system of non-interacting spins. Moreover, the third type originates from the action of an external magnetic field. The second type of paramagnetism occurs in a coupled spin system undergoing a magnetic ordering. Again as mentioned in the first chapter, the spin paramagnetism observed in the metallic systems can be of two kinds *viz.* (i) Pauli type and (ii) Stoner-Wohlfarth type; however, these will be discussed in the section on itinerant magnetism. Besides, the well-known spin-orbit coupling intrinsic to each atom or ion in the system also plays important role in deciding magnetism in insulators quite often and sometimes even in metals.

2.4   MODEL FOR FERROMAGNETISM

We investigate first the situation corresponding to two isolated Hydrogen atoms being brought close to each other. Each of the two atoms has a nucleus containing a proton and one electron belonging to the 1s orbital. As the atoms approach each other, the two electrons belonging to them come under the influence of mutual Coulomb repulsion. This new interaction coming into play between the electrons acts as a perturbation to the original non-interacting atoms. It can also be seen that the spin states of these two electrons are completely uncorrelated when the atoms are far apart. It is to be examined whether the perturbation affects the spin correlation of the electrons and if yes how. We carry out the perturbation calculation with the unperturbed electronic state as the anti-symmetrized 2-electron state, having both space coordinates and the spin coordinates. One can easily see that the unperturbed state is 4-fold degenerate. These degenerate states are the following:

$$\psi_1(r_1,r_2) = \frac{1}{2}\big(\phi_a(r_1)\phi_b(r_2) - \phi_a(r_2)\phi_b(r_1)\big)\big(\alpha(1)\beta(2) + \beta(1)\alpha(2)\big) \quad (2.5)$$

$$\psi_2(r_1,r_2) = \frac{1}{2^{0.5}}\big(\phi_a(r_1)\phi_b(r_2) - \phi_a(r_2)\phi_b(r_1)\big)\big(\alpha(1)\alpha(2)\big) \qquad (2.6)$$

$$\psi_3(r_1,r_2) = \frac{1}{2^{0.5}}\big(\phi_a(r_1)\phi_b(r_2) - \phi_a(r_2)\phi_b(r_1)\big)\big(\beta(1)\beta(2)\big) \qquad (2.7)$$

$$\psi_4(r_1,r_2) = \frac{1}{2}\big(\phi_a(r_1\phi_b(r_2) + \phi_a(r_2)\phi_b(r_1)\big)\alpha(1)\beta(2) - \beta(1)\alpha(2)\big) \quad (2.8)$$

where, a and b are the two Hydrogen atoms under consideration. The variables r_1 and r_2 are the spatial coordinates of the two electrons. The functions ϕ and ψ are the space parts of the individual electron wave function (atomic wave function) and the full two-electron wave function respectively. The variables α and β are the spin states "up" and "down" respectively for the individual electrons. It can be seen from the form of the above 4 states that the first 3 states represent the "spin parallel" or "triplet spin" states for the two electrons; whereas the last state is the "spin antiparallel" or "singlet spin" state for the two electrons. The three spin-triplet states with total $S = 1$, correspond to the states with $m_s = 0$, $m_s = 1$ and $m_s = -1$, respectively. The spin-singlet state corresponds to $S = 0$ and $m_s = 0$.

We now treat the inter-electron bare Coulomb interaction $V_c = \dfrac{e^2}{|r_1 - r_2|}$ in perturbation. We calculate the 1st order correction to the unperturbed energy. For this, we evaluate the expectation value of V_c in the 4 unperturbed degenerate states mentioned above viz. ψ_1, ψ_2, ψ_3, and ψ_4.

The calculation turns out to be quite interesting! First of all, it is to be noted that V_c operates only on the space part of the wave functions. Again, the integrals involving the space part are of two types viz. (i) direct term and (ii) exchange term. The direct term (or integral) is defined as:

$$K = \iint \phi_a^*(r_1)\phi_b^*(r_2)V_c\phi_a(r_1)\phi_b(r_2)\,dr_1 dr_2 \qquad (2.9)$$

Again, the Exchange term (or integral) is given by:

$$J = \int\int \phi_a^*(r_1)\phi_b^*(r_2)V_c\phi_a(r_2)\phi_b(r_1)dr_1dr_2 \qquad (2.10)$$

The matrix elements involving the spin parts are equal to unity because of the orthonormal property of the spin states. All these lead to the following results for the 1st order correction to the energy:

$$\Delta E_1 = k - J \qquad (2.11)$$

where, the subscript 1 refers to the state ψ_1. Similarly, for the other three spin-triplet states again we find that ΔE gives the same result viz. $k - J$. For the spin-singlet state ψ_4 however, we find that:

$$\Delta E_4 = k + J \qquad (2.12)$$

The above results may be summarized as:

$$\Delta E_t = k + J \qquad (2.13)$$

and

$$\Delta E_s = k + J \qquad (2.14)$$

Producing an energy gap of magnitude $2J$ between the singlet (denoted by subscript s) and the triplet (denoted by subscript t) states, as J can be shown to be positive, i.e., $J > 0$. Thus, we find that as a result of the introduction of the inter-electron Coulomb repulsion, the degeneracy of the unperturbed states gets lifted. The spin-triplet states experience a lowering of energy whereas; the spin-singlet state becomes higher in energy. Therefore, both the quantum nature of the electrons and the antisymmetric property of the wave function of spin ½ particles (electrons in this case) lead to tendencies for parallel alignment of spins, in the case of repulsive interaction, as the triplet state is lower in energy compared to the singlet state. This, in fact, is the microscopic origin of ferromagnetism in insulators.

Our next aim is to find an "effective spin model" which gives the same energy spectrum as above, for a pair of spin ½ particles. By inspection, one can see that a quantum Hamiltonian of the following form:

$$H_{spin} = 2J S_1 \cdot S_2 \qquad (2.15)$$

where, S_1 and S_2 are the spin operators corresponding to the two electrons and J, the original exchange integral, is called "exchange constant" for this model. The above Hamiltonian has energy eigenvalue spectra which show the same gap *viz.* $2J$ between the spin-singlet and the spin-triplet states. Thus, the overall effect of the Coulomb interaction on the spin correlations can be generated equivalently by the action of an effective pseudo-spin-spin interaction. The above quantum spin Hamiltonian is the well-known Heisenberg model for ferromagnetism. So far, we have introduced the model specifically for spin[1] case with the spin operator's representable as the Pauli matrices; however, the model can be generalized under certain conditions for higher spins as well. This generalized model is used for describing ferromagnetism in insulators containing atoms with large spin, like *EuO, EuS*, etc., having $S = \dfrac{7}{2}$.

The mechanism described above to explain the origin of ferromagnetism in insulators, is also known as "potential exchange." This was first proposed by Dirac and Heisenberg. The term potential signifies that the matrix element of the electrostatic Coulomb potential between the two-electron states with exchange process is the key player in generating ferromagnetism.

So far, we have presented the Heisenberg model involving spin operators corresponding to electrons only on two atoms. In order to describe ferromagnetism of macroscopic solids, however, we will have to extend this to a lattice. Thus, the Heisenberg spin model corresponding to a lattice can be written as:

$$H_{lattice}^{spin} = -\sum_{i,j} J_{i,j} S_i \cdot S_j \tag{2.16}$$

where, S_i and S_j are the spin operators corresponding to the electrons at the lattice sites i and j, respectively. The exchange constant J_{ij} is the original exchange integral involving two electrons belonging to the sites i and j. Usually, in most of the materials, J_{ij} is finite only for the sites i and j as nearest neighbors and next-nearest neighbors. For higher neighbors, J_{ij} decays very sharply.

The quantum mechanical nature of the spin operators is manifested quite strongly in the appearance of the spin raising and spin lowering operators *viz.* S^+ and S^- respectively in the Hamiltonian. This is evident from the following relation:

$$S_i \cdot S_j = S_i^z S_j^z + \frac{1}{2}\left(S_i^+ S_j^- + S_i^- S_j^+\right) \tag{2.17}$$

The first term in the right-hand side of the above equation is the well-known Ising contribution, which tries to align the spins at the two sites, for ferromagnetic coupling. The next two terms *viz.* the *XY* contribution however try to flip the spins and destabilize the magnetic alignment. This destabilizing effect is generally important at finite temperature, where some of the spins are misaligned. At zero temperature, however, the system being at the ground state, all the spins are perfectly aligned with the highest m_s values and therefore the spin-flip processes cannot take place. At finite temperature, the spin-flip process involving the misaligned spins, leads to the propagation of excitations in the magnetically ordered state, as we will see later.

For an isotropic Heisenberg model, as has been described above in the Eqn. (2.16), the coupling constant is the same for the coupling of the spin components along all the three directions. For an anisotropic Heisenberg model, however, the coupling constants are to be character-ized as J_{ij}^X, J_{ij}^Y and J_{ij}^Z corresponding to the coupling between the x, y, and z components respectively. The nature of the ground state, the ordered state, and the excitations depend very crucially on the relative magnitudes of these coupling constants and also on the spatial dimen-sion of the lattice, for a Heisenberg ferromagnet. The origin of these anisotropies lies mostly in the splitting of degenerate electronic energy levels due to the crystal fields, which will be elaborated later. For the isotropic case, the ground state is the state with parallel spins on all the lattice sites in a lattice of any spatial dimension. The low energy spin excitations are characterized as traveling waves called "spin waves" in three-dimensions. In lower dimensions, there are speculations about the existence of other kinds of excitations besides spin waves. In the anisotropic case, particularly in the lower dimensions, there are very clear indications of various types of novel excitation channels besides the usual spin waves, while the ground state is also quite complex sometimes. Moreover, in certain cases, the magnetically ordered phase does not exist above zero temperature. In particular, the *XY*-anisotropic and the Ising-anisotropic models have very distinct properties regarding the nature of ground states and the excitations. We will take up all these aspects systematically.

As mentioned before, the elementary excitations can be generated from the ground state by the action of the *XY* part of the Heisenberg Hamiltonian. We utilize this to construct and study the collective excitations for an isotropic Heisenberg ferromagnet.

In a semi-classical approach to the spin-wave theory, the equations of motion of the spin components are considered in the classical limit with a spin configuration corresponding to a very low temperature, approximating Heisenberg equations of motion for the quantum spin operators by suitable classical spin vectors. Assuming the direction of ordering to be along the positive *z*-direction, the *z* component of the spin vector is taken to be approximately time-independent. Then the equations of motion of the *x* and the *y* components are linearized. This leads to a propagating wave solution involving a precessional motion of the spin vectors about the *z*-direction at every lattice site. This wave is the spin-wave in semi-classical theory.

Mathematically, the original quantum mechanical Heisenberg equation of motion behind this process is:

$$\frac{dS_i(t)}{dt} = \frac{2\pi}{ih}\left[S_i(t), H\right] \tag{2.18}$$

This leads to the familiar equation governing the quantum spin dynamics;

$$\frac{dS_i(t)}{dt} = \frac{2\pi}{h}\sum_{ij} J_{ij} S_i(t) X S_j(t) \tag{2.19}$$

Detailed analysis of the semi-classical scheme applied to the above equation, its consequences as well as the analysis of the truly quantum case will be done when we explicitly deal with the spin dynamics of Heisenberg ferromagnet in the ordered phase. Let us now focus first on the static properties of the ordered phase itself.

2.5   ANALYSIS OF THE FERROMAGNETICALLY ORDERED PHASE

The ferromagnetic Heisenberg Hamiltonian is given by Eqn. (2.16) has a rotational invariance in the spin space, as is seen easily. Thus, the ground state has an infinite degeneracy. Nevertheless, in reality, the degeneracy is lifted and the system chooses the configuration with spins aligned along one particular direction, as the ground state. This phenomenon of "symmetry breaking" is often facilitated by the presence of an external magnetic field like earth's magnetic field or the dipolar field within the magnetic system itself. Thus in the theoretical treatment of the ordered phase for the determination of various properties, we start from a Hamiltonian which contains a term representing the coupling of the individual spins to an infinitesimal external magnetic field, besides the usual Heisenberg term. This vanishingly small Zeeman term breaks the symmetry and causes the direction of the spin alignment to be the direction of the magnetic field.

Our Hamiltonian thus becomes:

$$H = -\frac{1}{2}\sum_{ij} J_{ij} \mathbf{S}_i \cdot \mathbf{S}_j - \sum_i \mathbf{h}_i \cdot \mathbf{S}_i \qquad (2.20)$$

where, the factor ½ in the first term prevents the double counting and h_i denotes the tiny external field. We would like to subject this model to certain theoretical treatments to extract various useful physical quantities and properties in the ferromagnetically ordered phase. One of the most important quantities of interest is the spontaneous magnetization. In the simplest theoretical scheme called "mean field theory," the above quantity can be obtained in a very straightforward manner. In mean field treatment, an interacting Hamiltonian is effectively replaced with a non-interacting Hamiltonian such that all the physical properties of the old and the new Hamiltonians are the same. By applying mean field theory to our above Hamiltonian (see Eqn. (2.19)), we get:

$$H_{mf} = -\frac{1}{2}\sum_{ij} J_{ij} \mathbf{S}_i \cdot <\mathbf{S}_j> - \sum_i \mathbf{h}_i \cdot \mathbf{S}_i \qquad (2.21)$$

where, $<S>$ is the expectation value (more precisely thermodynamic average) of S, calculated with H_{mf} itself. This expectation value of the spin operator is also the magnetization vector in the ordered phase. The above scheme leads to a self-consistent equation for $<S>$, as we will see shortly. Moreover, this procedure also takes us to introduce a new effective field called "Weiss Molecular Field," denoted by H_i such that:

$$H_i = h_i + \sum_j J_{ij} <S_j> \tag{2.22}$$

This Weiss field is the mean field which couples to the spins. It has contributions from the external magnetic field as well as the internal exchange field, as we see from Eqn. (2.21). Now assuming the external field h_i to be uniform and independent of lattice site i and furthermore choosing the direction of h, i.e., the symmetry breaking direction to be z, we get:

$$H_i = h + \sum_j J_{ij} <S_j^z> \tag{2.23}$$

Generally, as stated before, the sum over j goes through the nearest neighbors only. We also assume this in our mean field calculation, for the time being.

The mean field Hamiltonian can be rewritten as:

$$H_{mf} = -\sum_i H_i S_i^z \tag{2.24}$$

Therefore, the self-consistent equation for the magnetization (along the z-direction) becomes:

$$<S_j^z> = Tr\rho S_j^z \tag{2.25}$$

Denoting the magnetization by M, we get from the above equation:

$$M = SB_s\left(S\beta(h + JM)\right) \tag{2.26}$$

where, S is the magnitude of the spin on each lattice site, J is the sum of J over all the nearest neighbor sites, β is as usual inverse temperature and B_S is "Brillouin Function." This function is defined as:

$$B_s(x) = \frac{2S+1}{2S}\coth\left(\frac{2S+1}{2S}\right) - \frac{1}{2S}\coth\left(\frac{x}{2S}\right) \tag{2.27}$$

The function B_S appears in the self-consistent equation for M because of the summation over the eigenstates of S_z operator needed for carrying out the trace, in evaluating the thermal average in Eqn. (2.24). It is interesting to note that for the case of $S = \dfrac{1}{2}$, the above Eqn. (2.25) reduces to:

$$M = S\tanh\left(S\beta\left(h + JM\right)\right) \tag{2.28}$$

In the limit when the external field becomes infinitesimal, the quantity M represents the "spontaneous magnetization." This quantity is also known as the "order parameter." The self-consistent equation for the spontaneous magnetization (to be denoted by M_s) becomes:

$$M_s = SBs\left(S\beta\,J\,M_s\right) \tag{2.29}$$

The above equation has interesting solutions. They can easily be obtained by graphical approach. For $T > T_c$, i.e., $\beta < \beta_c$, the equation has $M = 0$ as the only solution; T_c being some critical temperature parameter. At the other extreme in the temperature regime *viz* at $T = 0$, the spontaneous magnetization M_s attains the saturation value S. The critical temperature parameter T_c is determined from the graphical solution as well. It is given as:

$$T_c = \frac{JS(S+1)}{3} \tag{2.30}$$

The above expression for T_c in the mean field treatment is a very well-known expression and is widely used. The temperature regime above T_c is the paramagnetic region and that below T_c corresponds to ferro-magnetically ordered phase. The magnetically ordered region and the paramagnetic region are also known as "long-range ordered phase" and the "short-range ordered phase" respectively. The long-range ordering implies that the spins are perfectly aligned throughout the infinite lattice. In the short-range ordered phase, however, the alignment of the spins are only partial and that too extends only up to a finite separation distance between the lattice positions of the spins.

It is possible to calculate the spin response function of this spin model in a very straightforward manner, within the mean field theory. One of the most important quantities is the spin susceptibility. This is defined as the derivative of the induced spin magnetization with respect to the infinitesimal applied magnetic field, as given below:

$$X_{spin} = \frac{\delta <S_z>}{\delta h} \qquad (2.31)$$

In the limit h tending to zero. This can be evaluated with the help of Eqn. (2.25). However, in the high-temperature regime, there is a simpler way of determining χ. It is based on a result from generalized linear response theory, known as the "fluctuation-dissipation theorem." In simple terms, this theorem relates the imaginary part of the dynamical susceptibility, a complex quantity, to the space-time dependent fluctuation function. In the case of static response functions, this theorem takes a simpler form. The static susceptibility is a real quantity and at high temperature, it becomes directly proportional to the static fluctuation involving the relevant physical quantity at two different sites. Applying this result to our situation involving the static spin susceptibility and the static spin fluctuation, we arrive at the following schematic equation:

$$\chi = \beta \, [<SS> - <S><S>] \qquad (2.32)$$

In the above equation, the quantity within the bracket in the right-hand side is the two-spin fluctuation. In the case of uniform spin susceptibility arising as a response to a uniform magnetic field, the above equation can be more explicitly written as:

$$\chi \, (q = 0) = \beta \, [<S \, (q = 0)> - <S \, (q = 0)><S \, (q = 0)>] \qquad (2.33)$$

where, q is the wave-vector of the static response function. In the general case of response to a static but spatially varying magnetic field, the above equation corresponding to finite wave-vector dependent spin susceptibility becomes:

$$\chi \, (q) = \beta \, [<S(q) \, S(-q)> - <S(q)><S(-q)>] \qquad (2.34)$$

where, $S(q)$ is the lattice Fourier Transform of the site spin operators S_i's and is given by:

$$S(q) = \frac{1}{N^{0.5}} \sum_i S_i \exp(iq \cdot r_i) \qquad (2.35)$$

where, N is the total number of lattice sites. In the paramagnetic phase, this equation is further simplified as:

$$\chi(q) = \beta C_q \qquad (2.36)$$

where, C_q is the static q-space spin-spin correlation function given by:

$$C_q = < S(q) S(-q) \qquad (2.37)$$

We now examine various possible cases. To start with, we take an ideal paramagnet with non-interacting spins, i.e., $J = 0$. In this case, it can be seen easily by making use of the Eqns. (2.33)–(2.35), that $\chi_{spin}(q)$ for all q assume the magnitude $\frac{S(S+1)}{3T}$. This is because of the finite contribution from only the on-site spin correlations in the expression for C_q. Thus, we recover the well-known Curie's Law for spin susceptibility in an ideal paramagnet at high temperature. It should however be remarked that we could have also derived Curie's Law by making use of Eqns. (2.25) and (2.26) with $J = 0$ and by taking the low magnetic field and high temperature limit.

For an exchange coupled paramagnet *viz.* in the temperature regime above T_c for a ferromagnet, the above result gets slightly modified due to the presence of the internal exchange field in the form of the mean field. To be more explicit, our result for χ obtained in the previous paragraph now describes the system's response to the combined field of the external magnetic field and the mean field. Recalling the expression of this combined effective field H_i:

$$H_i = h_i + J M \qquad (2.38)$$

This implies that:

$$M = \chi_{IP} (h + J M) \qquad (2.39)$$

where, *IP* is the abbreviation for ideal paramagnet. However, the true spin susceptibility is given by Eqn. (2.29). Thus, the above equation leads to, by making use of expression for χ_{IP}:

$$\chi_{true} = \frac{\chi_{IP}}{1 - J\chi_{IP}} \qquad (2.40)$$

Incorporating the expression for χ_{IP} obtained in the earlier page, we arrive at the following expression for the susceptibility:

$$X_{true} = \frac{S(S+1)}{3(T - T_c)} \qquad (2.41)$$

where, T_c is Curie temperature, as defined earlier. The above expression for χ is known as Curie-Weiss Law for spin susceptibility, valid in the paramagnetic phase of a Heisenberg ferromagnet. The susceptibility exhibits divergence at $T = T_c$, as expected. The expression for susceptibility, we have just derived, is in fact that corresponding to a uniform magnetic field. For the wave-vector dependent spin susceptibility, we get a similar expression with T_c replaced by $(T_c)_q$. The parameter $(T_c)_q$ is defined in way similar to that for T_c, viz.

$$(T_c)_q = \frac{J(q)S(S+1)}{3} \qquad (2.42)$$

where, $J(q)$ is the spatial Fourier transform of the exchange coupling J_{ij}. It may be remarked that in the equation for the usual ferromagnetic Curie temperature T_c (see Eqn. (2.28)) the quantity J actually stands for $J(0)$. The temperature $(T_c)_q$ represents the ordering temperature corresponding to a SDW with wave vector q. Obviously, for a purely ferromagnetic exchange coupling $T_c > (T_c)_q$. When the exchange interaction is absent, T_c, and $(T_c)_q$ both vanish and the Curie-Weiss Law reduces to Curie's Law, as expected.

Let us now focus on the ordered phase. Making use of Eqns. (2.25) and (2.29), we get the following expression for χ_{spin}.

$$\chi_{spin} = \frac{\beta N S^2 B'_s}{1 - \beta J N S^2 B'_s} \qquad (2.43)$$

where, $B'_s(J)$ is the value of the derivative of the Brillouin function corresponding to the Eqn. (2.26), with respect to its argument in the limit h

tending to zero. Using this equation we can determine the longitudinal spin susceptibility in the ordered phase, to be denoted by $X_{spin}^{long}(T)$. Moreover, we can also determine Curie temperature T_c from the condition that the susceptibility diverges at this temperature. This leads to:

$$\beta_c JNS^2 B_s' = 1 \qquad (2.44)$$

It can be verified that the above expression for T_c, as well as that shown in the Eqn. (2.30), obtained from the treatment of the ordered phase and the earlier expression for T_c obtained from the paramagnetic phase viz. Eqn. (2.42) paramagnetic phase, are all consistent. The Eqns. (2.40) and (2.41) are essentially RPA equations (random phase approximation; to be discussed in more detail in Chapter 4) for the spin susceptibility in the disordered phase, with the numerator representing the susceptibility corresponding to an ideal paramagnet.

The spontaneous magnetization satisfies the Eqn. (2.29). It can be seen very easily that at $T = 0$, M_s attains the saturation value S. Again for T close to T_c, the careful analysis of this equation leads to the behavior M_s varying as $(T_c - T)^{0.5}$. The thermal variation of the spin-spin correlation length (ξ) is given by $(1 - T_c/T)^{-v}$ close to the transition temperature in the paramagnetic phase, where v is 0.5.

It is important to stress that all our discussions pertaining to the ordered phase applies to a three-dimensional isotropic Heisenberg ferromagnet. In lower dimensions, the rigorous statistical mechanical reasons involving thermal fluctuations prohibit any finite temperature phase transition to the long range ordered state, as was shown by Mermin and Wagner. There may however be exotic type of "phase transition" of quite unconventional in nature, in two-dimensional systems under certain conditions. This will be taken up later in this chapter.

Let us now turn our attention to the spin dynamics of a Heisenberg ferromagnet. Just like the statics, in dynamics too, the ordered phase and the disordered (exchange coupled paramagnetic) phase exhibit very different properties. As stated before, the Eqn. (2.18) is the basic equation guiding the spin dynamics. As is obvious, an ideal spin paramagnet does not exhibit any spin dynamics. The equation, however, is to be supplemented with the suitable boundary conditions or initial conditions. These conditions in turn depend very strongly upon the temperature and the nature of the phase. As all the observables like the dynamic spin correlations, will have to be obtained as an ensemble average, the statistical distribution of the spin configurations plays a very crucial role in deciding the nature of the spin dynamics. In particular, it is found that the long-range ordered phase occurring at temperatures below T_c can easily sustain propagating modes viz. the spin-wave modes; whereas the paramagnetic phase with short-range correlations seen in the temperature regime above T_c can only allow damped modes or heavily damped propagating modes. In the ordered phase itself- however, the spin waves get slightly damped and softened as temperature increases.

In order to achieve a full microscopic understanding for the above processes and the features in the spin dynamics of the Heisenberg model, we will have to start our analysis on the basis of the equation of motion for the dynamic two-spin correlation function itself. The Heisenberg equation of motion for the two-spin correlation generates various higher spin (like 4-spin, etc.), correlations besides the 2-spin correlation itself. If we factorize the 4-spin correlation functions suitably in terms of 2-spin correlation functions, we can close the problem. This leads to a "spin wave" solution and exhibits softening of these modes with temperature as well. To generate experimentally observed damping, i.e., the finite width of the spin-wave spectra, we will have to go beyond this linear spin-wave theory. This can be handled more efficiently by employing the Green's Function technique with the quantum magnon operators and carrying out higher-order perturbative calculations. A magnon represents a quantized spin wave.

Let us now discuss all these approaches for an analytical understanding of the features of the spin waves and the nature of the spin dynamics in general. We first start with the linear spin-wave theory at very low temperature, within the semi-classical framework.

As stated before, we approximate the quantum mechanical spin operator by a classical vector by the following prescription assuming large spin:

$$S_i^{op} = \frac{2\pi}{h}\left(S\left(S+1\right)\right)^{0.5} s_i \qquad (2.45)$$

where, s_i is a three-dimensional unit vector at the i-th lattice site and S is the magnitude of the spin. Thus the quantum spin operator corresponding to spin S is replaced by a classical vector of magnitude $(S(S+1))^{1/2}$. This enables us to rewrite Eqn. (2.18) as:

$$\frac{ds_i(t)}{dt} = t_0^{-1} \sum_{<ij>} s_i(t) X s_j(t) \qquad (2.46)$$

where, we have assumed only near-neighbor exchange interaction between the spins. The quantity t_0 is the natural time unit in the problem and is given by:

$$t_0^{-1} = \frac{2\pi}{h} (S(S+1))^{\frac{1}{2}} \times J \qquad (2.47)$$

We now break up the above equation for the dynamics of the unit spin vector into the corresponding equations for the three components. Furthermore, we consider the system to be at a temperature close to zero. This physical situation implies that the magnitude of the component along the direction of ordering gets only slightly reduced from unity and remains more or less constant in the dynamics. At the same time, the other two components now acquire small but finite values and show time evolution. These time evolution equations however can be linearized. These can be expressed as:

$$\frac{ds_i^z(t)}{dt} = 0 \qquad (2.48)$$

$$\frac{ds_i^x(t)}{dt} = t_0^{-1}(1-\epsilon)\sum_{<j>}\left[s_i^y(t) - s_j^y(t)\right] \qquad (2.49)$$

and

$$\frac{ds_i^y(t)}{dt} = t_0^{-1}(1-\epsilon)\sum_{<j>}\left[s_j^x(t) - s_i^x(t)\right] \qquad (2.50)$$

where, $(1 - \epsilon)$ is the magnitude of the time-independent z-component which is the solution to Eqn. (2.48); ϵ being much less than 1. We now seek plane wave solutions to the Eqns. (2.49) and (2.50) in the following form,

$$s_i^x(t) = A_x \exp(i(k \cdot r_i - w_k t)) \qquad (2.51)$$

and

$$s_i^y(t) = A_y \exp(i(\mathbf{k} \cdot \mathbf{r}_i - w_k t)) \qquad (2.52)$$

Plugging these solutions into the Eqns. (2.49) and (2.50), we get the following dispersion relation:

$$w_k = t_0^{-1}(1-\epsilon)\left(6 - 2\cos\left(k_x a\right) - 2\cos\left(k_y a\right) - 2\cos\left(k_z a\right)\right) \qquad (2.53)$$

for a three-dimensional cubic system with nearest-neighbor ferromagnetic exchange interaction, a being the lattice parameter. The above equation (Eqn. (2.52)) is the well-known spin-wave dispersion. In the long-wavelength (low-k limit), the dispersion relation takes the following approximate form:

$$\omega_k = Dk^2 \qquad (2.54)$$

where, $D = t_0^{-1}(1 - \varepsilon)a^2$ is called the spin wave stiffness constant. This quadratic dispersion in the long-wavelength limit is characteristic of the nearest neighbor isotropic Heisenberg ferromagnet. It mimicks a free particle-like behavior.

The above analysis is strictly valid at zero temperature. As the temperature increases, ε will also increase causing a further reduction in the magnitude of the ordering moment. This leads to a decrease in the spin stiffness constant and hence in the magnitude of the spin-wave frequency. Thus with an increase in temperature, we see a softening of the spin waves. This can be described more clearly by the finite temperature analysis involving the dynamics of the two-spin correlation function. Then carrying out a time-dependent mean-field theory like approach, one can systematically bring out the softening effect. The scheme is briefly as follows: the two-spin correlation function $C_{ij}(t)$ is considered and its equation of motion is written down in the semi-classical description. Then it is closed by making use of suitable decoupling of the four-spin correlation functions occurring in the right-hand side, by invoking the static correlation functions. There are two-spin correlations with different indices present on the right-hand side as well. Now the trick is to go to the k-space from the real space and express the equation of motion in the diagonalized form. Mathematically this can be expressed as:

$$\frac{d^2 C_q(t)}{dt^2} = w_q^2 C_q(t) \tag{2.55}$$

where, $C_q(t)$, the q-space dynamical two-spin correlation function, is defined as:

$$C_q(t) = \frac{1}{N} \sum_{ij} C_{ij}(t) \exp\left(iq \cdot r_{ij}\right) \tag{2.56}$$

The real-space dynamical two-spin correlation function $C_{ij}(t)$ is itself defined as:

$$C_{ij}(t) = <S_i(0) \cdot S_j(t)> \tag{2.57}$$

where, <> now represents thermal average, i.e., expectation value at finite temperature.

The solution to Eqn. (2.55) above is the oscillatory behavior of the q-space dynamical two-spin correlation function in time domain, displaying a spin-wave in the dynamics of the individual spins. The spin-wave frequency ω_q is now a function of the temperature-dependent static correlation functions and exhibits softening with temperature with weakening of the magnetic order and the static correlation (see for example, Lovesey, 1986).

The above approach has a serious limitation. It did not distinguish between the longitudinal and the transverse components and hence is not applicable to the ordered or the broken symmetry phase. It was more of a toy approach, more meaningful for describing the paramagnetic phase occurring above T_c. In order to handle the situation in the ordered phase more rigorously, we define spin correlation functions corresponding to both longitudinal components and transverse components and then consider the corresponding equations of motion. We now treat the problem totally quantum mechanically. Denoting the direction of ordering as z direction, the transverse and the longitudinal dynamic spin correlation functions are given by $\langle S_i^x(0)S_j^x(t)\rangle + \langle S_i^y(0)S_j^y(t)\rangle$ and $\langle S_i^z(0)S_j^z(t)\rangle$, respectively. The equations of motion of the transverse correlation functions can be derived from those of the operators $S^+(q)(t)$ and $S^-(q)(t)$ (see Lovesey, 1986).

The Heisenberg equation of motion for $S^+(q)(t)$ after linearization, i.e., after replacing operator S^z by the classical number S (the magnitude of the spin) takes the form:

$$\frac{dS^+(q)(t)}{dt} = \frac{2\pi}{ih} 2S(J(0) - J(q))S^+(q)(t) \tag{2.58}$$

This has the solution:

$$S^+(q)(t) = \exp(-iw_q t)S^+(q)(t) \tag{2.59}$$

where, ω_q is equal to $(J(0) - J(q))$ is the spin-wave frequency, as we have seen earlier. Thus the time evolution of the operator $S^+(q)(t)$ itself has a spin-wave nature. We now proceed to evaluate the correlation function for the transverse components.

From the definition of the thermal average value of the operator products and also the commutation relations involving $S^+(q)(0)$ and $S^-(q)(0)$, it can be shown that:

$$< S^-(q) S^+(q) > = 2SNn(\omega_q) \tag{2.60}$$

where, $n(\omega_q)$ turns out to be Bose distribution function and is given by:

$$n(w_q) = \frac{1}{\exp\left(\dfrac{hw_q\beta}{2\pi}\right) - 1} \tag{2.61}$$

Recalling the definitions of the operators S^+ and S^- one can easily evaluate the correlations involving the longitudinal components in the presence of a spin-wave with wave vector q:

$$< S_i^x(0)S_{i+r}^x(t) >= n(w_q)\frac{S}{N}\cos(q \cdot r - w_q t) \tag{2.62}$$

The y-component also obeys exactly the same equation as above. Furthermore, the xy correlations are governed by the following equation:

$$< S_i^x(0)S_{i+r}^y(t) + S_{i+r}^y(t)S_i^x(0) >= n(w_q)\frac{2S}{N}\sin(q\cdot R - w_q t) \quad (2.63)$$

In deriving the above two equations, we have assumed that $n(\omega_q)$ is much greater than 1. The presence of β in $n(\omega_q)$ implies that the transverse correlations are temperature dependent, as expected.

The longitudinal correlation function is rather trivial. By the very nature of the spin-wave excitation, the longitudinal correlation is time-independent. However, the magnitude of the longitudinal component is also temperature-dependent, as we will see later.

The above analysis takes us in a very natural way to the introduction of the concept of "magnon," as touched upon earlier. The identification of the spin-wave mode having a propagating vector q, with the time evolution of momentum-space dynamical quantum transverse spin correlation function corresponding to qth Fourier component helps us to quantize the spin waves in terms of the new quantum quasiparticle magnon.

The mathematical transformations defining the real space spin deviation operators corresponding to the magnons are known as Holstein-Primakoff transformations (HPT). They are given as:

$$S_i^z = S - b_i^+ b_i \quad (2.64)$$

$$S_i^+ = (2S)^{\frac{1}{2}}\left(1 - \frac{1}{2S}b_i^+ b_i\right)^{\frac{1}{2}} b_i \quad (2.65)$$

$$S_i^- = (2S)^{\frac{1}{2}} b_i^+ \left(1 - \frac{1}{2S}b_i^+ b_i\right)^{\frac{1}{2}} \quad (2.66)$$

where, b_i^+ and b_i are the local spin-flip (deviation) creation and destruction operators at site i, respectively. The operators b's obey Bosonic commutation algebra $viz.$

$$\left[b_i, b_i^+\right] = 1 \quad (2.67)$$

So that, the spin angular momentum algebra is given below:

$$\left[S_i^+, S_i^-\right] = 2S_i^z \tag{2.68}$$

and other relations involving the components of the spin operator are satisfied. For simplicity, we have dropped the 'hat' on the operators. Unlike the usual bosons, however, the operators b's will have to satisfy a constraint that $0 < b_i^+ b_i \leq 2S$. This is because of the underlying constraint that the spin deviation along any direction cannot exceed the magnitude $2S$. Using these spin deviation operators b's we now try to rewrite the ferromagnetic near-neighbor Heisenberg spin model. Performing a detailed algebra and keeping terms up to order of $1/S$, we arrive at the following form for the spin model:

$$H = S\sum_{<ij>} J_{ij} \left(b_i^+ b_i - b_i^+ b_j \right) \tag{2.69}$$

The above Hamiltonian can now be diagonalized by going over to the k-space and introducing magnon operators. The magnon operators are defined by the Fourier transformations of the spin deviation operators in real space b_i's. It is expressed mathematically as:

$$b_q^+ = \frac{1}{N^{\frac{1}{2}}} \sum_i b_i^+ \exp\left(-iq \cdot r_i\right) \tag{2.70}$$

and

$$b_q = \frac{1}{N^{\frac{1}{2}}} \sum_j b_j \exp\left(iq \cdot r_j\right) \tag{2.71}$$

The operators on the left-hand side of the above two equations *viz.* (2.70) and (2.71) are the creation and destruction operators respectively, corresponding to the magnon of wave vector q. Now it is quite straightforward to rewrite our Hamiltonian (see Eqn. (2.69)) in terms of the magnon excitations, as:

$$H = \text{constant} + \sum_q E_q b_q^+ b_q \tag{2.72}$$

where, E_q is the energy of a magnon with wave-vector q and is given by:

$$E_q = (J(0) - J(q))S \qquad (2.73)$$

which is exactly identical to the expression for classical spin-wave frequency ω_q.

It is quite easy to prove that the magnon operators also satisfy the bosonic commutation algebra; just like the real space spin deviation operators (see Eqn. (2.67)).

The introduction of the magnon operators also makes it possible to construct a one magnon state. It is defined as:

$$|1_q> = S_q^- |0> \qquad (2.74)$$

where, the operator S_q^- is defined as:

$$S_q^- = \sum_i S_i^- \exp(-iq \cdot r_i) \qquad (2.75)$$

Thus, a one-magnon state is a superposition of quantum states with unit spin deviation at various lattice sites, with a phase factor dependent upon the wave-vector of the magnon. It describes a propagating mode of a unit spin deviation through the lattice.

The Bosonic nature of the magnons can be exploited to calculate various properties like the specific heat and the spontaneous magnetization as functions of temperature in the ordered phase of a Heisenberg ferromagnet. Let us first look at the specific heat. The excitation energy of the ferromagnetic spin system in the ordered phase is given by:

$$E_{ex} = \sum_q \frac{hw_q}{2\pi} n(w_q) \qquad (2.76)$$

$n(\omega_q)$ is the usual Bose distribution function given by:

$$n(w_q) = \frac{1}{\exp\left(\dfrac{hw_q}{kT2\pi}\right) - 1} \qquad (2.77)$$

We now note the fact that the magnon energy has a quadratic dispersion in the long-wavelength limit, as has been shown earlier. We make use of this property and also go over from the lattice to the continuum limit. This leads to:

$$E_{ex} = (\text{constant}) \cdot \int_0^{q\,max} dq \alpha q^4 \exp\left(-\frac{\alpha q^2}{T}\right) \qquad (2.78)$$

where, α stands for the parameter $\hbar SJq^2a^2/2\pi$. It should be kept in mind that the above equation for E_{ex} has been written down for low temperature. As the upper limit in the integral, q_{max} is very large, we may regard the upper limit to be infinite. This simplification takes us to see that E_{ex} is proportional to $T^{\frac{5}{2}}$. Since the specific heat C_v is defined as $\frac{\delta E_{ex}}{\delta T}$, we immediately arrive at the result that C_v is proportional to $T^{\frac{3}{2}}$.

Similarly, one can calculate the temperature dependence of the spontaneous magnetization in the ordered phase. We have already shown that a magnon carries a loss of magnetization of one unit spin. Thus making use of the Bose distribution function, we can easily calculate the total loss of magnetization at any temperature. This in turn gives us the spontaneous magnetization as a function of temperature. More quantitatively, the spontaneous magnetization is given by:

$$M = S - m(T) \qquad (2.79)$$

where, $m(T)$ is the thermal loss of the magnetization due to the creation of magnons. It is given as:

$$m(T) = \frac{1}{N} \sum_q m_q(T) \qquad (2.80)$$

where, $m_q(T)$ is the amount of spin carried away by the magnons at temperature T. This is exactly equal to $n_q \cdot 1$. Thus it is quite easy to evaluate $m(T)$ by going over to the continuum limit, as is done earlier. This leads to the result that $m(T)$ is proportional to $T^{\frac{3}{2}}$. This is the well-known Bloch's

Law for the decay of spontaneous magnetization with temperature in the ordered phase of a three-dimensional isotropic Heisenberg ferromagnet.

It must be kept in mind, however, that both these results *viz.* those for the specific heat and the spontaneous magnetization, are valid at low temperature. At higher temperatures, the non-linearity neglected in rewriting the Hamiltonian in terms of the spin deviation operators b's becomes quite important and induces magnon-magnon interaction. Then magnons get damped and acquire a finite life-time. This makes the magnon distribution function deviate very strongly from the ideal Bose distribution function and the theoretical results given in the Eqns. (2.78) and (2.80) become invalid.

The approach described above with the diagonalized Hamiltonian does not predict the softening of magnons either. For both broadening and softening of magnon spectra, we will have to go beyond the non-interacting magnon Hamiltonian (given by Eqn. (2.72)), make use of suitable magnon Green's Function, and carry out a perturbation calculation.

In terms of the magnon operators, one defines the Green function $G_{q,q'}(t)$ of Zubarev type as (see for example Lovesey, 1986):

$$G_{q,q'}(t) = -i\theta(t) < \left[b_q(t), b_q^{'+}(0) \right] > \qquad (2.81)$$

where, $\theta(t)$ is the unit step function. One can take the Fourier transform of the above Green function with respect to time and get a new function $G_{q,q'}(\omega)$. The advantage of handling this new function is that its poles give the energies of the elementary excitations of the system described by the Hamiltonian.

Now, making use of the HPT as before, we do away with the low temperature and large S considerations and rewrite Heisenberg ferromagnetic spin Hamiltonian by keeping the non-linear terms as well:

$$H = \text{constant} + \sum_q S\left[J(0) - J(q) \right] b_q^+ b_q - \frac{1}{N} \sum_{q_1,q_2,q_3,q_4} b_{q_1}^+ b_{q_2}^+ b_{q_3} b_{q_4} \delta_{q_1+q_2+q_3+q_4} \left[J(q_1 - q_3) - J(q_3) \right]$$

$$(2.82)$$

The third term on the right-hand side is the new term representing the magnon-magnon interaction which we would like to treat perturbatively. Using this more realistic Hamiltonian, we now consider the equation of motion of the Green function $G_{q,q'}(t)$ and arrive at the following equation:

$$wG_{q,q'}(\omega) = \delta_{q,q'} + \frac{2\pi}{h} << \left[b_q, H \right]; b_{q'}^+ >> \tag{2.83}$$

where, the symbol $<< A; B >>$ represents the operation $<[A(t), B] >$, as is used in the Green function. It is clear that the Green function on the right-hand side of the above equation involves a Green function containing four Bose operators, besides the original Green function. The original Green function can be shown to have the following structure:

$$G_{q,q'}^0(\omega) = \delta_{q,q'} \frac{1}{\omega - \omega_q} \tag{2.84}$$

Thus, the unperturbed Green function has a pole at the non-interacting magnon frequency ω_q. The other Green function containing the four Bose operators when evaluated generates the "self energy" correction to the bare magnon mode. It has both real and imaginary parts. The real part can be shown to lead to softening of the magnon modes; whereas the imaginary part causes the damping of the magnons. The evaluation of this higher Green function is of course quite involved and need the study of its equation of motion followed by suitable decoupling schemes. The decoupling implies a factorization into a static correlation function involving these operators and this higher Green function itself. The static correlation function can be calculated using the standard commutator algebra. The higher Green function on the other hand, satisfies a self-consistent equation which can be solved analytically with some effort. All these can be utilized to cast the perturbed original Green function in the following form:

$$G_{q,q'}^0(\omega) = \frac{1}{\omega - \omega_q - \Sigma_q(\omega) \frac{2\pi}{h}} \tag{2.85}$$

where, Σ_q is the self-energy correction referred earlier. The detailed calculation shows that Σ is a complex quantity. The real part indicating the softening of the magnon mode is given by:

$$\Sigma^1_q(\omega) = -J\,A_q \sum_p n_p B_p \qquad (2.86)$$

where, A and B are two material–dependent parameters involving lattice structure, etc., and n_p is as usual the magnon occupation number. The parameters A and B are positive. Thus, Σ^1 is negative, indicating the softening of the magnon modes. Again, the imaginary part of the self energy, denoted by Σ^2 is given by:

$$\Sigma^1_q(\omega) = -JC \sum_{q1,q2} n_{q1} D_{q1,q2,q} \qquad (2.87)$$

where, C and D are again two material-dependent parameters, very much related to A and B. From these expressions for real and imaginary parts of self-energy, we can clearly see that both softening and damping of the magnon modes increase with temperature as the expectation value of the magnon occupation number (n) goes up with temperature.

Before we conclude the discussion about ferromagnetism and the static and dynamic properties of the ordered phase, let us make a few remarks regarding the phase above T_c. As pointed out earlier, the uniform static susceptibility obeys the Curie-Weiss Law. The spin dynamics is, however, more complex. The reason is that in this short-range ordered phase, the occupation number of magnons extrapolated from the ordered phase below T_c would be too high for finite order perturbation theory to work. To be rigorous, magnons are highly unstable because of the presence of multi-magnon processes in this phase. Thus, no spin wave-like dynamics is observed and the time evolution is predominantly diffusive in character. Nevertheless, for large values of $|q|$ viz. q close to the zone boundary, there is sometimes signature of a damped propagating character in the spin dynamics, as manifested by the occurrences of peaks at finite ω in the constant q-scan of the dynamical structure function $S(q, \omega)$, mimicking "spin-wave-like modes." This short length scale feature is sometimes interpreted as "precession of spins under local exchange field" [Shastry, Edwards, and Young; Lindgard; Lovesey; Chaudhury and Shastry; Demmel, Chatterjii, and Chaudhury; Boni and Shirane; Mook]. Rigorously speaking, however, the occurrences of both diffusive modes and damped propagating modes in different regions of q-space are consequences of the non-linearity of the coupled equations of motion for the spins. This conclusion is brought out by detailed theoretical analysis for the Heisenberg model combined with the experimental investigations for the materials *EuO* and *EuS*, which are ideal realizations for the 3D ferromagnetic Heisenberg model (Chaudhury and Shastry, 1988, 1989).

2.6   MODEL FOR ANTIFERROMAGNETISM

The phenomenon of antiferromagnetism is purely a quantum effect and has no classical analog. Nevertheless, amongst the magnetic materials, the antiferromagnets far outnumber the ferromagnets. For the insulating and semiconducting antiferromagnets, the origin lies in the existence of the strong on-site Coulomb repulsion amongst the electrons. More precisely, the phenomenon occurs in the half-filled band of the Mott-Hubbard insulators (MHIs). According to the conventional 1-electron band theory, a half-filled band implies a highly conducting metallic state, as has been briefly discussed in the first chapter. In these Mott-Hubbard systems, however, the on-site inter-electron repulsion is much larger than the electronic bandwidth and therefore the band theory treatment breaks down. In the absence of energetically feasible inter-site electronic hopping, the electrical conductivity vanishes. The virtual processes *viz.* hopping and hopping back involving two adjacent sites however are still possible and lead to antiferromagnetic correlations between electrons.

Let us now start from our model Hamiltonian and analyze the origin of antiferromagnetism mathematically.

The Mott-Hubbard systems are modeled by the well-known Hubbard model. This model is represented as:

$$H_{\text{Hubbard}} = t \sum_{<ij>,\sigma} c_{i\sigma}^{+} c_{j\sigma} + U \sum_{i} n_{i\sigma} n_{i-\sigma} \qquad (2.88)$$

where, t is the electronic hopping amplitude involving atomic orbitals belonging to two adjacent lattice sites, c, and c^{+} are the usual fermion destruction and creation operators, n is the fermion occupation number operator and U is the on-site Coulomb repulsion matrix element between the electronic states with opposite spins belonging to the same atomic orbital. The quantity U is also known as "Mott-Hubbard repulsion." It must be kept in mind that this model is an approximate version of the more general model of interacting electrons on the lattice, in the narrowband limit. For narrowband metals like *Fe*, *Ni*, etc., the bandwidth is much larger than *U*. However, the insulating oxides and semiconductors like *NiO*, *MnO* fall in the other region of the parameter space *viz.*, where *U* is very large compared to the bandwidth zt (z being the coordination number of the lattice). Moreover, in these oxides, the direct hopping between *Ni* sites or *Mn* sites is negligible; the hopping is predominantly through mixing with *O* orbitals. The Oxygen atom plays the role of a ligand which mediates the indirect hopping process. Interestingly both these oxides are also antiferromagnets. In fact, most of the Mott insulators turn out to be antiferromagnets as well. The exchange interaction originates from the Coulomb blockade, with the kinetic term being treated perturbatively. Let us now proceed systematically and extract an effective spin model for these Mott insulators.

We consider the situation with 1 electron per lattice site in a particular atomic orbital. Let the lattice be of *N* sites. Therefore, the ground state of Hubbard model, in the absence of the *t*-term, is 2^N-fold degenerate, which implies all singly occupied sites with arbitrary spin projections of either $+\frac{1}{2}$ or $-\frac{1}{2}$ are contained in this state. Thus, the ground states have no spin correlations. Let us now treat the hopping term within the first-order perturbation scheme. This gives:

$$\Delta E^1_{mn} = t \sum_{<i,j>,\sigma} c^+_{i\sigma} c_{j\sigma} <G_m||G_n> \qquad (2.89)$$

where, $|G_m>$ and $|G_n>$ are any two states belonging to the manifold of the degenerate ground states. Now it can be seen very easily that the action of the *t*-term on $|G_n>$ takes it to an excited state containing a doubly occupied site. Thus, the overlap of this newly formed state with $|G_m>$ is zero. Therefore, the first-order correction to the energy ΔE^1 vanishes.

Next, we try out the second-order perturbation correction. The correction corresponding to the 2nd order process involving the pair of adjacent lattice sites *i* and *j* are given by:

$$\Delta E_{mn}^2 = \sum_{|EX>} \frac{H_{mn}^{EX}}{E_G - E_{EX}} \qquad (2.90)$$

where, $|EX>$ is an excited state vector of the system; E_G and E_{EX} are the energies corresponding to the ground and the excited states of the system. The matrix element H_{mn} is given by:

$$H_{mn}^{EX} = <G_m|t \sum_{<ij>,\sigma} c_{i\sigma}^+ c_{j\sigma}|EX><EX|t \sum_{<rs>,\sigma'} c_{r\sigma'}^+ c_{s\sigma'}|G_n> \qquad (2.91)$$

The detailed and systematic investigation of the above expressions leads to:

$$\Delta E_{mn}^2 \frac{1}{N} = \frac{t^2}{U} \qquad (2.92)$$

where, N, and z are the total number of lattice sites and the coordination number respectively, provided $|G_n>$ has Neel type of spin arrangement and $|G_m>$ is identical to $|G_n>$ or differ by just 1 spin-pair exchange. The Neel state is defined as the state where all the adjacent pair of spins is anti-parallel. If however, the above conditions are not satisfied by the states $|G_n>$ and $|G_m>$, then ΔE^2_{mn} vanishes. It is interesting to observe that the 2nd order process is able to lift the degeneracy of the unperturbed ground state and it is energetically preferable to have ground states with Neel-like correlations.

These considerations guide us to construct the following effective pseudo-spin model of Heisenberg type:

$$H_{Heisenberg} = -\sum_{<ij>} J_{ij} S_i \cdot S_j \qquad (2.93)$$

where, the nearest neighbor exchange constant $J_{<ij>} = J$ is given by:

$$J = -\frac{t^2}{U} \qquad (2.94)$$

The negative magnitude of J indicates an antiferromagnetic coupling between the spins of the electrons belonging to the nearest neighbor sites. Thus, this Hamiltonian induces a tendency for anti-parallel alignment, i.e., Neel-like correlations, of the electronic spins belonging to the nearest neighbor sites. However, the quantum mechanical nature of the spin operators also causes a quantum flipping of the spins, and thus Neel state is fragile. We will take up this aspect shortly and have an elaborate discussion on the nature of the ground state of Heisenberg antiferromagnet.

To generate an effective Heisenberg spin model, which has the same energy eigenvalues as the parental electronic model in the large U-limit, we need a more rigorous treatment with the operators. First of all, the electronic operators involved in the 2nd order processes will have to be expressed as equivalent spin operators. For spin ½ particles, this can be done by making use of Schwinger-Dyson transformation viz.

$$S_i^+ = C_{i\uparrow}^+ C_{i\downarrow} \tag{2.95}$$

$$S_i^- = C_{i\downarrow}^+ C_{i\uparrow} \tag{2.96}$$

$$S_i^z = \frac{1}{2}\left(n_{i\uparrow} - n_{i\downarrow}\right) \tag{2.97}$$

provided, the number of particles per site is 1, i.e.,

$$n_{i\uparrow} + n_{i\downarrow} = 1 \tag{2.98}$$

The above transformation maps the 2nd order processes as the operation under the following equivalent Hamiltonian:

$$H = -\sum_{<ij>} J_{i,j}\left(S_i \cdot S_j - \frac{1}{4}\right) \tag{2.99}$$

The energy expectation values of the above equivalent effective quantum spin Hamiltonian, to be denoted by H_{QHAF} from now onwards, are different for the spin antiparallel, and spin parallel states as expected. For the Neel state, the energy per spin pair is $\frac{-J}{2}$ and for state representing a ferromagnetic ground state the energy per pair is zero, indicating clearly that anti-parallel alignment of spins is preferred. Thus in the insulating limit of the half-filled band, the Hubbard model leads to antiferromagnetism.

The quantum mechanical nature of the spin operators occurring in H_{QHAF} also has a very interesting consequence. It brings out the existence of quantum fluctuations and this has a tremendous influence on the nature of the ground state. To elaborate on this point, the operators S^+ and S^-, i.e., the spin raising and spin lowering operators respectively, flip the near neighbor spins easily in anti-parallel configuration and thus destabilize the Neel state. The operator product involving the S_z's however try to strengthen the lifetime of the Neel state. This competition is very severe in low-dimensional systems and as a result, the ground state is drastically different from Neel state in one dimension. The exact ground state corresponding to H_{QHAF} is known only for spatial lattice in 1-dimension. This was done by H. Bethe in 1932. The ground state looks very complicated. An approximate and simplified version of this is a state constructed as a product of spin-singlet states, as pointed out by L. Pauling and P. W. Anderson (Fazekas and Anderson, 1972; Majumdar and Ghosh, 1969). For a three-dimensional lattice, however, the Neel state has a very large lifetime and for all practical purposes serves as a very good ground state. For antiferromagnets on 2D lattice, however, the situation is somewhere in between; the ground state has a very large Neel component but there are also "domain wall-like" configurations with moderate weightages present in the ground state.

With the nature of the ground state having a Neel-like structure for a 3d QHAF, the antiferromagnetically ordered phase at finite temperature also displays a similar structure consisting of two sub-lattices. This is known as the "bi-partite" magnetic lattice structure. It consists of an "up spin sub-lattice" and a "down spin sub-lattice." Analogous to a 3d ferromagnet, here too both the sub-lattices exhibit spontaneous magnetization of equal magnitude but of opposite directions, below a certain temperature. This temperature is known as "Neel temperature (T_N)," similar to Curie temperature for a ferromagnet. The analysis of the static properties in the ordered phase can again be done in a way parallel to that employed for a ferromagnet. The spontaneous sub-lattice magnetization can be studied theoretically with the mean-field technique. This also brings out a way to estimate the T_N. The spin dynamics can also be studied by introducing magnons with the help of Holstein-Primakoff transformations. In contrast to the ferromagnet, however, the Weiss field assumes two different values on the two sub-lattices, when mean-field theory is applied to study the ordered phase. Again, the bi-partite lattice structure shows up in the spin dynamics of a 3d antiferromagnet as well, with the appearance of two distinct types of magnons. Moreover, many of the properties of the 3d antiferromagnet can be recovered from the corresponding properties of a 3d ferromagnet by performing the staggering transformation, i.e., the wave vector q going over to $q + (\pi, \pi, \pi)$. We now demonstrate some of the above features of a 3d QHAF by presenting brief mathematical calculations.

Let us look at the situation when the nearest neighbor exchange coupling J in much larger than the tiny externally applied field. Our Hamiltonian then becomes:

$$H = -\sum_{<ij>} J_{ij} S_i \cdot S_j - \sum_t h_i \cdot S_i \qquad (2.100)$$

where, the site i and site j belong to two different sub-lattices A and B respectively in the first term and belong to the same sub-lattice in the second term. In the absence of the external field, the sub-lattice magnetizations are the spontaneous magnetizations M_A and M_B produced along the z-direction and satisfying:

$$M_A = -M_B = M_s \qquad (2.101)$$

In the presence of the applied field along the z-direction, however, the magnitude of M_A becomes larger than M_B producing a net magnetization. The Weiss fields corresponding to the two sub-lattices are given by:

$$H_i^A = JM_B + h_i \qquad (2.102)$$

and

$$H_i^B = JM_a + h_i \qquad (2.103)$$

We can further assume h_i to be uniform and independent of the lattice site. Then the magnetizations satisfy the equations:

$$M_A = SB_s (S\beta (h + JM_B)) \qquad (2.104)$$

and

$$M_B = SB_s \left(S\beta \left(h + JM_A \right) \right) \tag{2.105}$$

These coupled equations will have to be solved self-consistently. The more interesting case appears when the external field tends to zero and we deal with the spontaneous magnetizations. Then the above set of equations takes a very simple form *viz.*

$$M_s = SB_s \left(-S\beta \, JM_s \right) \tag{2.106}$$

The above equation looks exactly identical to the corresponding equation for the Heisenberg ferromagnet, provided we make the transformation J goes to $-J$. This has the natural consequence in the expression for the antiferromagnetic ordering temperature T_N. Comparing with that for the ferromagnetic ordering temperature T_C, derived earlier, we get:

$$T_N = -JS(S+1)^{\frac{1}{3}} \tag{2.107}$$

keeping in mind that J is negative for antiferromagnets.

Similarly, in the paramagnetic phase of an antiferromagnet, i.e., for T above T_N, we can show that the spin susceptibility obeys the following relation:

$$\chi_{AF} \left(q = 0 \right) = \frac{S(S+1)}{3(T+T_N)} \tag{2.108}$$

Making use of the self-consistent field approach as used in the case of a ferromagnet earlier and also using the relation between T_N and J. For finite wave vector (q) however, the susceptibility expression is different, as expected. For $q = (\pi, \pi, \pi)$ it takes an interesting form *viz.*

$$\chi_{AF} \left(q = \pi, \pi, \pi \right) = \frac{S(S+1)}{3(T-T_N)} \tag{2.109}$$

Thus susceptibility of an antiferromagnet for the staggered wave vector diverges at T_N, exactly in a way exhibited by a ferromagnet for $q = 0$ mode at T_C.

The spin response discussed above deals with the situation when the applied field is along the same direction as the broken symmetry direction *viz.* z-direction. One may also look at the transverse spin response, i.e., with the magnetic field applied transverse to the broken symmetry direction. In the ordered phase of a ferromagnet, it can be argued that the transverse spin susceptibility in the long-wavelength limit diverges; whereas in the case of an antiferromagnet it can be shown to be a constant. In the paramagnetic phases of both of these types of systems, however, the transverse responses become identical to the respective longitudinal ones, as expected in the non-broken symmetry phase.

The spin dynamics of QHAF is very fascinating and intriguing indeed. For lower spatial dimensions, there are signatures and conjectures regarding the existence of topological spin excitations. They are distinctly different from the spin waves or magnon modes. We however confine our discussions to the spin waves only. We start our analysis in a way very similar to that used for a 3d ferromagnet.

We introduce the Bosonic operators a's and b's for the creation and destruction of the spin deviations corresponding to the up and down sub-lattices, respectively. Thus, the HPT take the following form for the bi-partite lattice:

$$S_i^z = S - a_i^+ a_i \tag{2.110}$$

$$S_i^+ = (2S)^{\frac{1}{2}} \left(1 - \frac{a_i^+ a_i}{2S} \right)^{\frac{1}{2}} a_i \tag{2.111}$$

$$S_i^- = (2S)^{\frac{1}{2}} a_i^+ \left(1 - \frac{a_i^+ a_i}{2S} \right)^{\frac{1}{2}} \tag{2.112}$$

for the up spin sub-lattice. Similarly, for the down spin sub-lattice we have:

$$S_i^z = -S + b_i^+ b_i \tag{2.113}$$

$$S_i^+ = (2S)^{\frac{1}{2}} \left(1 - \frac{b_i^+ b_i}{2S} \right)^{\frac{1}{2}} b_i^+ \tag{2.114}$$

$$S_i^- = (2S)^{\frac{1}{2}} b_i \left(1 - \frac{b_i^+ b_i}{2S}\right)^{\frac{1}{2}}$$

(2.115)

As in the case of a ferromagnet, we can go from the real space to the k-space to construct the corresponding magnon operator's *viz.* a_K's and b_k's. Substituting the spin operators by these magnon operators in the Hamiltonian given by Eqn. (2.99), we arrive at a form for H which is non-diagonal in both a and b. We can get a diagonalized form for H by performing the standard Bogoliubov-Velatin (BV) Transformation to generate new sets of bosonic operators α's and β's. The transformation looks:

$$\alpha_K = u_k a_k - v_k b_k^+$$

(2.116)

$$\beta_K = u_k b_k - v_k a_k^+$$

(2.117)

The transformation coefficients u_k and v_k can be chosen to be real but satisfying the constraint:

$$u_k^2 - v_k^2 = 1$$

(2.118)

In terms of these new bosonic operators, our Hamiltonian can be rewritten in the linearized approximation as:

$$H = -2NJS(S+1) + \sum_k \frac{h}{2\pi} w_k \left(\alpha_k^+ \alpha_k + \beta_k^+ \beta_k + 1\right)$$

(2.119)

where, ω_k is the magnon frequency for the antiferromagnet. The interesting contrast with the corresponding diagonalized form for the ferromagnetic model is the contribution of zero-point magnon energy to the ground state energy of the QHAF. This is a manifestation of the fact that the Neel state is not a true ground state of the QHAF, as it is not even an eigenstate of the Hamiltonian, whereas it has been used as the starting state for the construction of the magnon operators. In other words, the Neel state itself contains zero point or virtual magnons, although the small quantum fluctuations around this state have been considered to generate these magnons.

The dispersion relation of these antiferromagnetic magnons for a simple cubic crystal is given by:

$$w_k = 2J\left(1 - \gamma_k^2\right)^{\frac{1}{2}} \qquad\qquad (2.120)$$

where, the lattice structure-dependent parameter γ_k is given by:

$$\gamma_k = \left(\cos(k_x a) + \cos(k_y a) + \cos(k_z a)\right)^{\frac{1}{3}} \qquad\qquad (2.121)$$

For low values of k, i.e., in the long-wavelength limit, the above equations lead to ω_k as proportional to k. This dispersion is like sound wave dispersion, in contrast to the case of a ferromagnet where it is like that of a free particle. The two magnon branches are degenerate; however, in the presence of an external magnetic field, this degeneracy is lifted.

The detailed spin dynamics of a 3d antiferromagnet can be studied in the same way as has been done for a ferromagnet and all the effects like softening, damping of the antiferromagnetic magnons can be analyzed in great detail. The effect of these magnons on the temperature dependencies of various quantities like specific heat, magnetization, etc., is also worked out in a similar fashion.

Another new class of magnetic materials very similar to the anti-ferro-magnets is "ferrimagnets." They are a special type of antiferromagnets with unequal spins on the two sub-lattices. As a result, they exhibit a net magnetization even in the absence of an external magnetic field. Moreover, the magnon branches are distinct and split into an acoustic branch and an optical branch. Some of the Fe-based systems known as "ferrites" belong to this class.

A very novel phenomenon occurs when in the same material; both ferromagnetic and anti-ferromagnetic interactions are present. This leads to an effect known as "frustration" in the spin system sometimes. It implies that the spins do not have a clear cut preference between the aligned and the anti-aligned configuration. This also gives rise to a huge degeneracy of the magnetic ground state.

Besides the above possibilities, even pure antiferromagnetic interactions between both nearest and next-nearest neighbors on a square lattice, can also lead to frustration. Moreover, pure antiferromagnetic nearest neighbor interaction on a triangular lattice also produces a large degree of frustration.

The above processes involving competition between different types of interaction amongst the spins may sometimes create a new phase of magnetic ordering called "spin glass." The term "glass" refers to the absence of spatial translational symmetry but at the same time, it retains a frozen correlation of the spins over a very large macroscopic time duration. It is distinct from a paramagnet in that it has a very high degree of spin correlation persisting theoretically over infinite time; whereas in a paramagnet the spin correlations decay very sharply with time in an exponential fashion. Various alloys made out of Cu, Fe, Mn, Co, Ni, etc., in proper stoichiometric ratios, are found to have spin-glass phases in different temperature regimes. Many of these alloys exhibit a variety of phase transitions between the paramagnetic, ferromagnetic, antiferromagnetic, and spin glass phases, as the temperature is lowered. The spin-glass freezing temperature T_g is an important crossover temperature below which the spin freezing takes place with the absence of spatial ordering. At this temperature, the spin susceptibility exhibits a cusp. The order parameter for a spin glass is of two types $viz.$ (i) Edwards-Anderson and (ii) Sherington-Kirkpatrick. Both of these involve spin auto-correlation functions for an infinite time interval.

Before we close our discussions on the magnetic ordering in insulators, it is very important to point out that in many of the real magnetic materials, where orbital effects and crystal field splittings are strong; there exists another type of spin-spin interaction known as Dzyaloshinskii-Moriya (DM) interaction. The form of this interaction is $D_{ij} \cdot (S_i X S_j)$, where D_{ij} is DM coupling function which is anti-symmetric under the exchange of i and j. This interaction essentially originates from spin-orbit coupling and can induce many interesting properties like 'canting of spins' rather than alignment or anti-alignment of spins. Besides other types of direct and indirect exchanges, DM interaction also needs to be taken into account in explaining the magnetic properties of many of these real systems. These processes can lead to a new phenomenon known as "spin Hall effect" in which the application of an external electric field or charge current causes spin current to develop, as discussed in Chapter 1. In magnetic semiconductors, there are also other exciting phenomena such as 'photo-magnetism' and formation of exotic quasi-particles such as 'ferrons'(Nagaev, 1983). The former is a magnetism induced by laser irradiation causing the transition of electrons from nearly filled valence band to empty conduction band enabling manifestation of magnetic behavior. On the other hand, the composite particle ferron made up of a fermion surrounded by Bosonic magnons, is formed during hopping of electrons or holes through a magnetically ordered lattice. This process is very similar to the formation of quasi-particle 'polaron' in an ionic lattice (to be discussed in the later chapters).

2.7   ITINERANT MAGNETISM

The spin models we have considered so far *viz*. the Heisenberg models describe magnetism arising from localized electrons, i.e., mostly the *f*-electrons. Many of the well-known magnetic materials however have itinerant or band electrons coming from *s* or *d*-bands which are responsible for magnetic phenomena occurring in them. The ferromagnetism in transition metals like *Fe, Ni, Co*, antiferromagnetism in *Cr*, and paramagnetism in alkali metals and in *Pd* owe their origin to the itinerant electrons. Again, it is the Coulomb interaction that plays a key role in the generation of the exchange interaction in these systems. This process is however opposed by the kinetic energy of these band electrons. It is this competition that decides the nature of magnetism here.

We will mostly confine our discussions to paramagnetism of both Pauli-type and of exchange enhanced Stoner type. We will also briefly touch upon some of the models predicting long-range ordering involving ferromagnetism and anti-ferromagnetism.

2.7.1   PAULI PARAMAGNETISM

The most fundamental and elementary form of itinerant magnetism is Pauli paramagnetism. This effect occurs for a non-interacting system of band electrons or holes. The origin of this paramagnetism is the Pauli Exclusion Principle and the Fermi distribution function. The essential physics is the following: in the absence of an external magnetic field, the fermi spheres corresponding to both up spin and down spin are of equal size and it produces zero magnetization. In the presence of a very small magnetic field, however, the degeneracy of the up spin and down spin electrons are lifted and the fermi sphere corresponding to the electrons with spin orientation (up) parallel to the direction of the magnetic field gets bigger than that corresponding to the down spin. This results in a net magnetization in the direction of the applied field. Moreover, the spin susceptibility at zero temperature turns out to be proportional to the electronic density of states (EDOS) at the Fermi energy.

For simplicity, we approximate the band electrons by a non-interacting Fermi sea. Under the application of an external magnetic field H along the z-direction, the total number of electrons with up and down spins is given respectively as:

$$n_\uparrow = \frac{1}{N} \sum_k n_{k\uparrow} = \int_{-\mu_B H}^{E_F} N(E)dE \qquad (2.122)$$

and

$$n_\downarrow = \int_{\mu_B H}^{E_F} N(E)dE \qquad (2.123)$$

where, $N(E)$ is the EDOS and E_F is the Fermi energy. The upper limit of the integration has been put at E_F, as the analysis is being carried out at zero temperature and that the Zeeman energy is very small compared to the Fermi energy for tiny external field. In the integrals of the above equations, we can make the following changes of variables viz. E' denoting $E + \mu_B H$ for the first equation and E' representing $E - \mu_B H$ for the second equation for convenience. The magnetization proportional to the difference between the up spin density and the down spin density, then simply becomes $2mu_B^2 HN(0)$ with $N(E)$ approximated as the density of states at the Fermi energy denoted by $N(0)$ and brought outside the integral sign for small magnitude of magnetic field in the integrals with the new variables. This leads to the following expression for the linear spin susceptibility:

$$\chi_s = 2\mu B^2 N(0) \qquad (2.124)$$

This susceptibility is also known as "Pauli susceptibility" and will be denoted by χ_p. Even at finite temperature, for $T << T_F$, T_F being the Fermi temperature, the above expression holds good for the spin susceptibility of the non-interacting band electrons, exhibiting a temperature-independent magnetic response.

The spin response of most of the alkali metals exhibits Pauli susceptibility. However, for transition metals, the spin response is more complicated, as the band electrons (d-band electrons) are quite strongly interacting. Thus, we need modifications of our approach to take care of the many-body effects. The lowest correction is due to RPA, as has been introduced earlier in the context of the Curie-Weiss Law for Heisenberg systems.

To incorporate the RPA corrections to χ_p, we for simplicity consider a Hubbard model in the weak coupling regime, i.e., for $U \ll t$. At the level of mean-field theory, the Hubbard Hamiltonian with nearest-neighbor hopping can be approximated as:

$$H_{Hubbard}^{MF} = \sum_{<ij>} t_{ij} c_{i\sigma}^+ c_{j\sigma} + U \sum_i <n_{i\uparrow}> n_{i\downarrow} + U \sum n_{i\uparrow} <n_{i\downarrow}> \qquad (2.125)$$

Let $<n_{i\uparrow}>$ and $<n_{i\downarrow}>$ be site independent and be denoted by $n_\uparrow$ and $n_\downarrow$ respectively. Then the net spin moment is given by:

$$M = \mu B (n_\uparrow - n_\downarrow) \qquad (2.126)$$

The expressions for $n_\uparrow$ and $n_\downarrow$ are very similar to the corresponding ones for the non-interacting case in the presence of the magnetic field H, excepting the changes in the lower limits of the integral. The lower limits are changed to $-\mu_B H + U n_\downarrow$ and $\mu_B H + U n_\uparrow$ for the up spin electron and the down spin electron, respectively. The calculation exactly similar to that in the non-interacting case then leads to:

$$\chi_s = \frac{Xp}{1 - N(0)U} \qquad (2.127)$$

The above expression for spin susceptibility for the interacting band electrons is known as Stoner's formula. It clearly shows that the suscep-tibility is enhanced by the inter-electron Coulomb interaction modeled by Hubbard U. It is expected as the mean-field generated from the Hubbard term, acts like a Weiss exchange field. The enhancement in the paramagnetic spin susceptibility, as shown by Stoner's formula, is seen very prominently for Pd which never becomes a ferromagnet. Again, the enhancement of the EDOS at the Fermi level itself can also boost the spin susceptibility, even with rather a weak Coulomb interaction, as seen in some of the heavy fermion systems (Steglich et al., 1984).

Moving over from itinerant or band paramagnetism, we now briefly discuss itinerant ferromagnetism and anti-ferromagnetism. Though this problem is a very tricky one, involving both band structure and many-body effects, we present a simple description first in terms of a one-band Hubbard model in the weak coupling limit. Again, there are also detailed many-body treatments involving interacting electron gas in high, low, and intermediate-density limits, which provide quite valuable insight towards the microscopic origin of itinerant ferromagnetism and itinerant antiferromagnetism or SDW. Besides, we also touch upon the two band scenario corresponding to semi-metal or semiconductors for itinerant magnetism of long-range type, involving time-dependent Hartree-Fock approach.

2.7.2   ITINERANT FERROMAGNETISM

In the Stoner formula introduced in the last section, we saw the enhancement in the uniform spin susceptibility brought about by the Coulomb interaction. In fact, it signals a tendency towards ferromagnetic ordering although it is not achievable generally, because of the existence of the weak coupling condition *viz.* $N(0)U \ll 1$. In certain cases, however, the small value of U (allowing the RPA to be valid) may be overcompensated by the existence of a large value of $N(0)$ and the denominator diverges. This signals the onset of ferromagnetism. In other words, the Coulomb interaction may induce a long-range magnetic ordering depending upon the symmetry of the electronic band or the Fermi surface. In particular, the mean-field decoupling of the interaction term involving the band states in a specific way leads to ferromagnetism, antiferromagnetism or even SDW. We now take up first the occurrence of ferromagnetism briefly.

We consider a Hubbard model for a half-filled band in the small U limit. Then the Hubbard term is approximated by its mean-field contributions involving the band states with zero momentum transfer. These particle-hole pairs of zero momentum can cause a splitting between the spin up and spin down sub-bands with the magnetization proportional to U. The more detailed description is that an energy gap develops at the original Fermi energy position between the two sub-bands and the magnitude of the energy gap is directly proportional to the magnetization itself. This is brought out by the diagonalization scheme of BV type, applied onto the mean-field approximated Hubbard model. The system however as a result of the emergence of this energy gap, undergoes a metal-insulator transition. Thus, the itinerant ferromagnetism of the kind described above, does not retain the metallic character of the system, and converts it to a band insulator! The other main drawback of this approach is that it predicts magnetic phase transitions at finite temperature even in $1d$ and $2d$.

Another scenario of itinerant ferromagnetism is through a systematic perturbation theoretic approach for interacting electron gas. At high electron density, the kinetic energy dominates over the Coulomb interaction effects and the system is spin paramagnetic. However, as the electron density is lowered, the magnetic correlations are enhanced, and in a certain density regime, the magnetic ordering sets in. Although the nature of the ordering is not completely settled, the onset of ferromagnetic ordering is a possibility.

The more complete description of realistic itinerant ferromagnetism is through a two-band approach. The interaction is again coulombic but a generalized one and the bands may be electron or hole-like. This is applicable mostly to semiconductors and semi-metals with different concentrations of electrons and holes. The generalized time-dependent Hartree-Fock approach yields an electron-hole pairing of triplet nature, which produces a long-range ferromagnetic ordering of true itinerant nature (Ginzburg and Kirzhnits, 1982).

A point to be noted is that ironically in spite of the vast richness in the physics of itinerant magnetism, the ferromagnetism in transition metals is often attempted to be described by the Heisenberg model! This leads to several paradoxes which are collectively known as the Iron problem (Antropov, 2003; Shastry et al., 1979).

2.7.3   ITINERANT ANTIFERROMAGNETISM

Just like the emergence of ferromagnetism, the phenomenon of antiferromagnetism can also be studied within the framework of the Hubbard model in the weak coupling regime. Again, the Hartree-Fock approach with the order parameter set at staggered magnetization with a two-sublattice structure, on the basis of itinerant electron spins, yields an energy gap. The metallic system turns into an insulator via processes involving the mixing of the states in the upper and lower subband, differing by the wave vector (π, π, π) and generating an antiferromagnetically long-range ordered state. This type of antiferromagnetism is known as Slater Band Antiferromagnetism. The energy gap is again proportional to the sub-lattice magnetization. The Fermi surface will of course have to obey a certain symmetry called "nesting" for this antiferromagnetic order parameter to assume a finite value.

The alternative approach is through a generalized two-band model, as was done for the ferromagnetism. The electron-hole singlet pairing gives rise to itinerant antiferromagnetism sometimes even SDW. The detailed band structure enters into the calculations here and contributes to the determination of the magnitude of the wave vector of the SDW (Ginzburg and Kirzhnits HTSC, 1982).

The itinerant magnetic order parameters are decided by the expectation values of the relevant quantities, which essentially are local spin operator components. Calculations of these take care of the hopping effects of the itinerant electrons through the quantum mechanical treatment of the Hamiltonian itself.

The spin excitations in itinerant ferromagnets and anti-ferromagnets are quite novel and are quite distinct from those of the localized systems. There are contributions from both transverse and longitudinal spin components here unlike the case of the localized systems (Moriya, 1985). Moreover, these collective modes of magnonic nature undergo Stoner damping for large wave-vectors, in analogy with Landau damping experienced by collective charge density oscillations characterized by plasmons (Lovesey, 1986).

2.7.4 MAGNETISM IN NON-MAGNETIC METALS BY MAGNETIC IMPURITY

An interesting effect is observed when magnetic impurities are incorporated in the non-magnetic metals. In particular, the situation involving a rare earth impurity in a transition metal host is an extremely interesting system for both theoretical and experimental studies. This leads to a variety of phenomena in condensed matter physics. The most notable of them are (i) mixed valency, (ii) heavy fermion physics, (iii) superconductivity, and (iv) magnetism. Here we discuss the magnetism aspect only.

The fundamental Hamiltonian describing this type of system is known as the Anderson model. It involves the Hamiltonian corresponding to the itinerant and conduction electrons from the s band of the transition metal and that corresponding to the atomic-like d or f electrons of the rare earth element. Mathematically it is given as:

$$H_{And.} = \sum_{k\sigma} \epsilon_k \, C^+_{k\tilde{A}} C_{k\tilde{A}} + \epsilon_d \sum_{m\sigma} d^+_{m\sigma} d_{m\sigma} + \sum_{km\sigma} V_{km} \left(C^+_{k\sigma} d_{m\sigma} + h.c. \right) + \sum_{m\sigma} U_{dm} n_\sigma n_{-\sigma}$$

$$(2.128)$$

where, the Coulomb interaction in the form of Hubbard-term for the localized d or f electrons, has been included in the Hamiltonian. The term containing the coefficient V is the hybridization term involving the matrix element of the single electron kinetic energy operator between the band states and the localized atomic-like states. The Hubbard term is generally very large and is greater than the site energies and the hybridization term. The relative strengths of the parameters in the above Hamiltonian decide the various effects and phenomena observed in these systems.

One way of treating this problem is to perform a Schrieffer-Wolf transformation (SWT) on Anderson Hamiltonian and eliminate a few terms to arrive at an effective Hamiltonian in the lowest order, known as the Kondo model (Nagaev, 1983; Mahan, 1981). This model is an important one for describing the magnetism in these systems and is given by:

$$H_{Kondo} = H_{sd} + H_{s-electrons} + H_{site\ energy\ of\ local\ electrons} \qquad (2.129)$$

where, the first term on the right-hand side is given by:

$$H_{sd} = -\sum_{kk'} J_{kk'} \left[S_d^z \left(C^+_{k\uparrow} C_{k'\uparrow} - C^+_{k\downarrow} C_{k'\downarrow} \right) + S_d^+ C^+_{k\downarrow} C_{k'\uparrow} + S_d^- C^+_{k\uparrow} C_{k'\downarrow} \right] \quad (2.130)$$

This exchange interaction between the itinerant electrons and the localized electrons at the impurity site is antiferromagnetic in nature (J is negative) and is local. The second and the third term describe the kinetic energy of the band electrons and the single site energy of the localized electrons at the rare earth impurity site. These two terms occur also in the Anderson model discussed earlier. There are various possible effects associated with these models. The most notable amongst them are: formation of "Kondo singlets," occurrence of "Kondo minimum" for dc electrical resistance, and the phenomenon of mixed valency. The antiferromagnetic interaction causes a conduction (band) electron to form a bound spin anti-parallel configuration with the rare-earth atomic electron, which may be a stable state. On the other hand, the remaining conduction electrons sometimes can also suffer spin-flip scattering from this bound object, causing the electrical resistance to suddenly start increasing with the decrease in temperature below a certain temperature. Thus, this model is important from both the magnetic and transport properties of these materials. Moreover, the hybridization between the rare earth impurity-level electronic states and the conduction band states of the transition metal causes the number of localized electrons to fluctuate, leading to the phenomenon of mixed valency or valence fluctuation.

The magnetism in terms of a long-range magnetic ordering can take place in these systems, when a large number of rare earth impurity atoms are introduced. The effective spin-spin interaction between two rare earth impurity atoms can be obtained by retaining the higher-order terms beyond the Kondo model, in the SWT carried out as mentioned earlier. This leads to the well-known"RKKY" interaction (developed first by Rudolf, Kittel, Kasuya, and Yosida) between the localized rare earth spins, when averaging over the band electronic states are performed.

The RKKY interaction looks formally like a Heisenberg interaction; although the dependence of the interaction on the spatial separation between the rare-earth atoms is very non-trivial. It is an oscillatory function combined with power-law decay with the distance between the rare-earth atoms. Thus, the interaction changes character from ferromagnetic to antiferromagnetic and vice versa, and the strength also changes, as the separation is varied. This leads to competition amongst ferromagnetic and antiferromagnetic couplings and can cause different types of magnetic orderings. The emergence of the oscillatory behavior depends very strongly on the existence of the sharp Fermi surface of the itinerant electrons. Thus with an increase in temperature, as the Fermi distribution function is smeared out, the oscillatory component in the RKKY interaction also gets weakened.

A very interesting situation arises when the rare earth impurities form a lattice. The starting Hamiltonian then takes the form of Anderson lattice Hamiltonian. The consequences are then more dramatic. The system may show a crossover between a long-range ordering corresponding to localized magnetism of antiferromagnetic type to itinerant paramagnetism of Stoner type, as the temperature is lowered. Moreover, the itinerant phase is often characterized as a "heavy fermi-liquid" which even turns superconducting at very low temperatures. In some of the materials, however, the itinerant magnetism may be of long-range anti-ferromagnetic or SDW type coexisting with superconductivity. Some of these effects will be dealt with in the chapter on superconductivity and on various effective theories about the normal phase.

2.7.5 VAN VLECK PARAMAGNETISM

A very different kind of paramagnetism arises when the external magnetic field induces virtual transitions between the non-magnetic ground state and an excited state with a net orbital angular momentum (implying a finite orbital magnetic moment) of an atom or ion. This leads to a temperature-independent magnetic susceptibility of positive magnitude, due to the 2nd order process at low temperature. Mathematically, this is brought out by the relation:

$$X = -\frac{\delta^2}{\delta H^2}\Delta E^2 \qquad (2.131)$$

where;

$$\Delta E^2 = [<GS|\mu_B L_z H|EX><EX|\mu_B L_z H|GS>]\,\frac{1}{E_{GS}-E_{EX}} \qquad (2.132)$$

This is known as Van Vleck paramagnetism. This operates in the solid, as the collection of atoms also exhibits this effect (White, 2007; Kittel, 1953). The above analysis can easily be extended to include the spin magnetic moment for the full paramagnetic response as well.

2.7.6 SPIN DYNAMICS IN THE PARAMAGNETIC PHASE OF FERROMAGNET AND ANTI-FERROMAGNET CORRESPONDING TO LOCALIZED MAGNETISM

While we developed, a perturbative framework for studying spin dynamics in the long-range ordered phase of a ferromagnet or antiferromagnet, the corresponding problem in the paramagnetic phases of the respective systems, known also as "exchange-coupled paramagnets," is very difficult to handle. Some of the many-body techniques based on Mori's continued fraction expansion involving 2-pole and 3-pole truncations, have been applied but have received limited success. On the other hand, non-perturbative techniques like Monte Carlo-molecular dynamics (MCMD) simulation have been quite successful in both qualitative and quantitative understanding of the spin dynamics in the paramagnetic phase and have been in good agreement with the experimental results. The new inelastic neutron scattering experiments using polarized neutron beam, had provided substantial challenge and motivation to the theoreticians for the investigation of spin dynamics in this phase (Chaudhury and Shastry, 1988, 1989; Boni and Shirane, 1986; Mook, 1981; Chatterji, Demmel, and Chaudhury, 2006).

2.7.7 MAGNETISM IN LOW-DIMENSIONAL SYSTEMS

So far, the discussion has been confined to the three-dimensional systems where there occurs a long-range magnetic ordering below a finite temperature. The situation is however quite different in the case of low-dimensional systems. In particular, the quasi-1d and quasi-2d localized magnetic systems have provided very novel concepts and issues for the condensed matter physicists to investigate. The nature of excitations is quite different here from those in the three-dimensional cases. In such low dimensional magnetic systems *viz.* 1d, quasi-1d, and 2d, the excitations are often of topological nature. The magnetic phase transition to a long-range ordered state at finite temperature is also absent here in the conventional sense; although a topological phase transition known as Berezinskii, Kosterlitz, and Thouless (BKT) transition, may occur for two-dimensional ferromagnetic systems in some cases *viz.* for XY-anisotropic models. This unconventional phase transition involves the unbinding of topological excitations occurring in the form of vortex-antivortex pairs at some finite temperature called T_{BKT}. This leads to the appearance of a "central peak" in the plot of $S(q, \omega)$ versus ω in the constant q scan. The problem is even more complex in the truly quantum regime, i.e., with low spin situations *viz.* $S = \frac{1}{2}$, as seen in the inelastic neutron scattering experiments performed on K_2CuO_4, with the appearance of a central peak for q near zone boundary as well (Chaudhury and Paul, 2003; Sarkar, Chaudhury, and Paul, 2012). Motivated by the success of the proposed phenomenology to the above ferromagnetic material, an extension of the idea of BKT transition to XY-anisotropic anti-ferromagnetic models in quantum domains on 2d lattice, has also been attempted later. This extended phenomenology was then applied to the experimental results from spin $\frac{1}{2}$ anti-ferromagnetic, insulating cuprates with very interesting consequences (Sarkar, Chaudhury, and Paul, 2016). Furthermore, possible transformation and conversion of multi-magnon modes to the quantum topological excitations in the case of 2d ferromagnets, have also been shown theoretically (Sarkar, Chaudhury, and Paul, 2015; Mattis, 1981) (Figures 2.1 and 2.2).

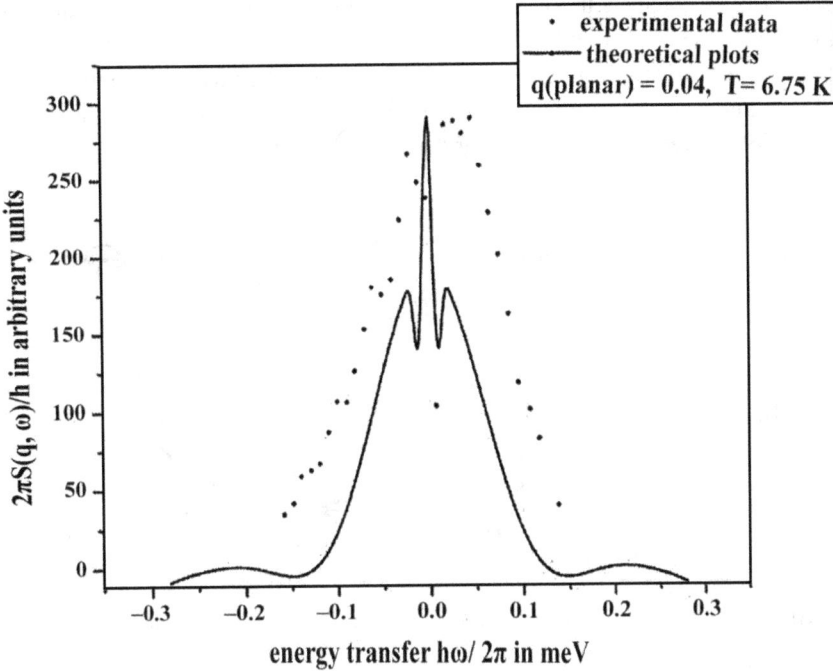

FIGURE 2.1 Dynamical structure-function vs. frequency curve exhibiting a central peak.

Source: Reprinted from the paper by Sarkar, S., Paul, S. K., & Chaudhury, R., (2012). EPJB. With kind permission of The European Physical Journal/Springer. © 2012).

For 1d antiferromagnetic systems, more precisely for nearest neighbor anti-ferromagnetic spin ¹isotropic Heisenberg chain, the ground state is very different from the well-known Neel state and looks quite complicated, as was first shown by Bethe through the detailed mathematical calculation in 1931. The low-lying excitations are of a solitonic type known as Clauxius-Pearson (CP) mode, but very interestingly, they have a dispersion relation very similar to that of an anti-ferromagnetic magnon. An approximation to this ground state is provided by a spin liquid or valence bond state (Majumder and Ghosh, 1969; Anderson et al., 1987). Furthermore, a very clear distinction between the behavior of the half-integral spins and integral spins for the nearest neighbor anti-ferromagnetic spin chains was conjectured by Affleck and Haldane in 1987. According to them, the half-integral spin systems would exhibit gapless excitations whereas, the integral spin systems would show massive modes. This prediction was later verified by experiments and was also established theoretically quite rigorously by the field theoretical calculations (Fradkin and Stone, 1988). Regarding 1d ferromagnets, a lot of results are available from the work of several theoreticians, with an indication of the existence of solitonic excitations (Luther and Peshchel, 1975; Fisher et al., 2009; Reiter, 1981; Mikeska and Steiner, 1991).

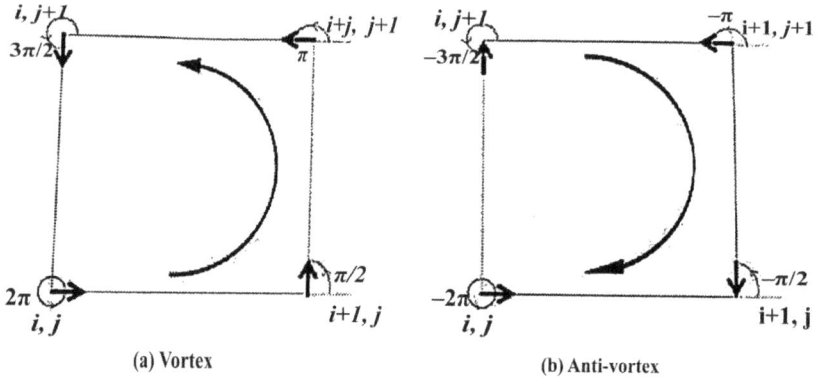

(a) Vortex

(b) Anti-vortex

FIGURE 2.2 Vortex-anti-vortex configurations in 2d XY-anisotropic quantum spin models.

Source: Reprinted with permission from the paper by Sarkar, S., Chaudhury, R., & Paul, S. K., (2015). IJMPB.© 1987, 2015. World Scientific Publishing Co., Inc.).

These path-breaking contributions of Kosterlitz, Thouless, and Haldane for low dimensional magnetism, have been duly recognized in the form of Nobel Prize in Physics for the trio in the year 2016.

The other challenging areas of research in magnetism comprise of investigation of the nature of ordering in nano and mesoscopic systems. Even though the magnetic orderings here are of the short-range type in the true sense, they have tremendous technological application possibilities. In fact, the nano and mesoscopic magnetic systems in low dimensions are of paramount interest to researchers, both experimentalists and theoreticians. Besides, there are important research areas like spintronics, Heusler alloys, spin current and spin Hall angle, magneto-caloric effects, exchange bias (corresponding to hysteresis loops) for ferromagnets, etc., some of which have already been discussed briefly in Chapter 1.

2.7.8   EXPERIMENTAL TECHNIQUES

There are various experimental techniques to study magnetism in solids. The most common amongst them are Neutron Diffraction, Inelastic Neutron Scattering, NMR, ESR, perturbed angular correlation (PAC), Mossbauer spectroscopy, Muon Spin Rotation (μ SR), AC susceptibility measuring instrument like vibrating sample magnetometer (VSM), DC magnetization measuring instrument making use of magneto-optic Kerr effect (MOKE), etc. The existence of magnetic ordering and its nature with a detailed magnetic structure are verified and determined by neutron diffraction. The nature and dispersion of magnetic excitations and collective spin dynamics are studied with the help of an inelastic neutron scattering technique. The discovery and the wide use of polarized neutron beam in the last two decades, has been very useful to filter out the phonon contributions and extract the genuine magnetic scattering part. This provides information about the magnetic dynamical structure function $S(q, \omega)$. The other experimental techniques based on local probes like NMR, ESR, PAC, μSR, and Mossbauer spectroscopy determine the temporal evolution of spin at a local site, expressed by the spin auto-correlation function. The bulk static and uniform magnetic susceptibility is determined by the very last technique in the above list (See for details Atland and Simons, 2006; Dattagupta, 1987).

KEYWORDS

- **Berezinskii, Kosterlitz, and Thouless**
- **Dzyaloshinskii-Moriya**
- **Monte Carlo-molecular dynamics**
- **Neel temperature**
- **random phase approximation**
- **spin density wave**

REFERENCES

Affleck, I., & Haldane, F. D. M., (1987). *Phys. Rev. B., 36*, 5291.

Ashcroft, N. W., & Mermin, N. D., (2017). *Solid State Physics.* Brooks Cole/Cengage Learning India.

Atland, A., & Simons, B., (2006). *Condensed Matter Field Theory.* Chapter 5, Cambridge University Press, Cambridge.

Berezinskii, V. L., (1971). *Sov. Phys. JETP, 32,* 493.

Bethe, H., (1931). *Z. Phys., 71,* 205.

Boni, P., & Shirane, G., (1986). *Phys. Rev. B., 33,* 3012.

Chatterji, T., Demmel, F., & Chaudhury, R., (2006). *Physica B., 385,* 428.

Chaudhury, R., & Paul, S. K., (2013). *Adv. in Condensed Matter Physics.* Art. ID: 783420.

Chaudhury, R., & Shastry, B. S., (1988). *Phys. Rev. B., 37,* 5216; *Phys. Rev. B., 40,* 5036 (1989).

Chaudhury, R., (2006). *Jour. Magn. and Magn. Mat., 307,* 99.

Chaudhury, R., Demmel F., & Chatterji, T., (2011). *Dynamical Response of Single Bi-Layer Spin Model: A Theoretical Analysis.* arXiv: 1104.4197v1.

Costa, B. V., & Lima, A. B., (2012). *Jour. Magn. and Magn. Mat., 324,* 1999.

Costa, B. V., Gouvea, M. E., & Pires, A. S. T., (1992). *Phys. Lett. A., 165,* 179.

Dattagupta, S., (1987). *Relaxation Phenomena in Condensed Matter Physics.* Academic Press.

Des Cloizeaux, J., & Pearson, J. J., (1962). *Phys. Rev., 128,* 2131.

Fazekas, P., & Anderson, P. W., (1974). *Phil. Mag., 30,* 423.

Fradkin, E., & Stone, M., (1988). *Phys. Rev. B., 38,* 7215.

Fradkin, E., (1991). *Field Theories of Condensed Matter Systems.* Addison-Wesley, CA.

Haldane, F. D. M., (1982). *Phys. Rev. B., 25,* 4925.

Kittel, C., (1953). *Introduction to Solid State Physics.* Wiley, New York.

Kosterlitz, J. M., & Thouless, D. J., (1973). *J. Phys. C., 6,* 1181.

Lifshits, I. M., (1982). *Quantum Theory of Solids.* MIR Publishers, Moscow.

Lindgard, P. A., (1982). *J. Appl. Phys., 53,* 1861; *Phys. Rev. B., 27,* 2980 (1983).

Lovesey, S. W., (1986). *Theory of Neutron Scattering From Condensed Matter* (Vol. 2). Clarendon Press, Oxford.

Luther, A., & Peschel, I., (1975). *Phys. Rev. B., 12,* 3906.

Mahan, G. D., (1981). *Many Particle Physics.* Plenum Press.

Majumdar, C. K., & Ghosh, D. K., (1969). *J. Math. Phys., 10,* 1388.

Mattis, D. C., (1981). *The Theory of Magnetism I, Statics, and Dynamics.* Chapter 5, Springer.

Mikeska, H. J., & Steiner, M., (1991). *Adv. Phys., 40,* 191.

Mook, H. A., (1981). *Phys. Rev. Lett., 46,* 508.

Moriya, T.,(1985). *Spin Fluctuations in Itinerant Electron Magnetism.* Springer-Verlag, Berlin, Heidelberg.

Nagaev, E. L., (1983). *Physics of Magnetic Semiconductors.* MIR Publishers, Moscow.

Reiter, G., (1981). *Phys. Rev. Lett., 46,* 202.

Sarkar, S., Chaudhury, R., & Paul, S. K., (2015). *Int. J. of Mod. Phys. B., 29*(29), 1550209.

Sarkar, S., Paul, S. K., & Chaudhury, R., (2012). *Eur. Phys. J. B., 85,* 380.

Sarkar, S., Paul S. K., & Chaudhury, R., (2017). *Jour. Magn. and Magn. Mat., 421,* 207.

Shastry, B. S., Edwards, D. M., & Young, A. P., (1981). *J. Phys. C., 14,* L665.

Sheng, D. N., Motrunich, O. I., Fisher, M. P. A., (2009). *Phys. Rev. B., 79,* 205112.

White, R. M., (2007). *Quantum Theory of Magnetism, Magnetic Properties of Materials.* Springer Series in Solid-State Sciences.

Yamada, K., (2004). *Electron Correlation in Metals.* Cambridge University Press, Cambridge.

CHAPTER 3

Lattice Dynamics, Phonons, and Some Applications for Optical Properties of Solids

3.1 INTRODUCTION

A large part of condensed matter physics, both theoretical and experimental, deals with studies of crystal lattices, directly or indirectly. In particular, both crystal structure (i.e., lattice geometry) and the lattice dynamics can play sometimes very crucial roles in deciding the properties of various condensed matter systems and also the occurrences of different phenomena like ferroelectricity, superconductivity, and magnetism. There are also extremely interesting and exciting phenomena in solids involving various 'optoelectronic,' i.e., occurring with the combination of optical and electronic components processes. The occurrences of these processes in low-dimensional systems have found tremendous technological applications.

As explained detail in Chapter 1, the lattice of atoms or more precisely the ion cores provide the Bloch potential essential for the electronic energy bands to form in a regular solid. The ion cores themselves interact with each other to form a stable and well-defined lattice. This inter-atomic or inter-ionic potential is attractive at large spatial separation and repulsive at shorter distances. This is often modeled by Van-der-Waals or Lennard-Jones type of potential functions (Huang, 1975). Both of these functions exhibit the required spatial dependence of the inter-atomic potential. They arise from the interactions involving the nuclei and the electron clouds of the neighboring atoms. They have contributions from dipole-dipole and also higher-order multipole-multipole interactions in general. The attractive-repulsive nature of this inter-atomic interaction leads to the existence of an equilibrium distance between the near neighbor atoms. This is identified with the lattice parameter of the corresponding crystal lattice. In terms of detailed microscopics, all these processes involve overlap between the electronic wave-functions of the neighboring atoms.

A real crystal lattice however experiences the influence of thermal energy. This leads to lattice excitations of collective type and manifests as the lattice vibrations. They arise as the solutions of the coupled equations of motion for the atoms belonging to all the lattice sites. An atom displaced from its designated equilibrium position exerts a force on its neighbors, and causes a motion involving all the atoms of the lattice eventually. The classical treatment leads to the following equations of motion:

$$M_\alpha \frac{d^2 s_{n\alpha i}}{dt^2} = -K\left(2_{s n\alpha i} - s_{(n+1)\alpha i} - s_{(n-1)\alpha i}\right) \tag{3.1}$$

where, $s_{n\alpha i}$ is the displacement of the ion of the αi-th variety at the nth site and K is the force constant for the lattice dynamics.

If we take a plane wave solution to the above equations of motion *viz.*

$$s_n(t) = s_0(0)\exp(ikna - \omega_k t) \tag{3.2}$$

Then we readily discover a relation between the frequency and the wave-vector of this Fourier mode (ω_k, k) as given by:

$$\omega_k = (4K/M^{1/2})\sin\left(\left|\frac{ka}{2}\right|\right) \tag{3.3}$$

The above simple modeling of lattice dynamics is known as the Born-von-Karmann model (Kittel, 1953). In the long-wavelength limit, we recover the sound wave mode *viz.* ω_k proportional to k, as expected. Furthermore, we also see the Goldstone mode, i.e., the phonon frequency tends to zero, as the wave vector tends to zero.

At this juncture, it is worthwhile to make a comparison of the above features with the results and pictures obtained from the Debye model appropriate to the continuum (Kittel, 1953).

The Debye model assumes that there exists a maximum value of the wave-vector k_D up to which the linear dispersion of the lattice dynamics representing the acoustic mode, holds. The frequency corresponding to this maximum wave-vector is known as Debye frequency (ω_D). This cutoff wave-vector k_D is found by the condition that the number of degrees of freedom (number of atoms per unit cell times the number of unit cells) is equal to the number of k-vectors (modes). A very simple calculation based on this gives:

$$k_D = \left\{(6\pi^2) \times n\right\}^{1/3} \tag{3.4}$$

where, n is the number density of atoms. We now generalize the above scheme and go over to the normal mode description, to have the individual equations of motion rewritten in terms of the collective coordinates consisting of both position and momenta variables X_k and P_k, respectively. They are defined as the Fourier Transforms of the conjugate variables s_n and p_n, respectively. This enables us to cast the equations of motion as that of a "collective mode." The Hamiltonian expressed in terms of this normal mode or collective mode is given by:

$$H_{lattice} = \sum_k \left[\frac{1}{2M} P_k P_{-k} + \frac{M}{2} \omega_k^2 X_k X_{-k} \right] \tag{3.5}$$

The above equation is a manifestation of the collective oscillation of the lattice atoms, representing a collective mode of the lattice dynamics. Each individual atom participates in this collective motion.

In the case of the Einstein model, the phonon is assumed to be of a single frequency and dispersion-less (Kittel, 1953). We will discuss about this in somewhat details a little later.

There are a lot of applications of this study to real materials ranging from metals to ionic and covalent insulators. The mathematical form of the inter-atomic interaction $W(|r_i - r_j|)$ is very much dependent on the nature of the material and plays very important roles in deciding the dispersion relation of the lattice vibration. This potential W is modeled often by Van-der Waals interaction or by Lennard-Jones potential, to name a few (Huang, 1975). The lattice vibrations are responsible for the various characteristic features observed in electrical transport and electrical polarization properties of these materials. This is brought out clearly in the quantum mechanical treatment of the lattice vibrations and its interaction with the electrons.

The other important related aspect is the binding energy of a solid. In an ionic crystal of insulating nature, this binding or cohesion is due to the inter-ionic interaction. This can be calculated even classically, by taking into account the attractive interaction between the oppositely charged ions as well as the repulsive interaction between the ions of the same charge, all situated on the well-defined sites of the lattice. The total attractive interaction turns out to be of higher magnitude than the repulsive one, because of the ionic distances under the lattice geometry. This leads to the very important concept of "Madelung constant" (Madelung, 1996). It is defined as the constant of proportionality involved in relating the net binding energy of the ionic crystal to the nearest neighbor attractive Coulomb energy. These calculations are done classically, as mentioned earlier. To express mathematically,

$$E_{cohensive} = -A \frac{e^2}{4\pi \in_0 R} \tag{3.6}$$

where, A is Madelung constant, R is the nearest neighbor ionic separation and $\in_0$ is the background static dielectric constant. For binary lattices, Madelung constant generally lies between 1.5 and 2.0. For ternary and higher lattices, it may assume much higher values.

The calculation of binding energy of a covalent or a molecular crystal is however much more complex. It is to be carried out quantum mechanically using bonding and anti-bonding-like molecular orbitals as basis states. Besides, the electronic correlations will also have to be incorporated in these calculations for realistic descriptions of these types of solids. Often it is based on quantum many-body techniques like "total energy calculation," "configuration integral," and "coupled-cluster technique." These techniques are widely used in theoretical and computational quantum chemistry (Bartlett, 1997; Cox, 2010).

Regarding the metallic solids, the cohesive energy can be calculated by a very well defined quantum mechanical procedure based on electron gas perturbation theory (to be discussed at length in Chapter 4) for alkali metals. For other kinds of metals, a combination of density functional theory involving nontrivial exchange-correlation effects with band structure calculations are generally used (to be discussed in Chapter 4). For transition metals this calculation is even more formidable.

So far, our treatment of lattice dynamics has been purely classical. We now proceed with the quantization scheme to be applied on the normal modes. This causes the lattice vibrations to be quantized in terms of new quasi-particles called "phonons." The procedure briefly is as follows: from the collective coordinates P_k and Q_k we define two new quantities a_k and $a_k{}^+$ as a linear combination of the above collective coordinates. Mathematically it is described as:

$$a_k = A\left(Q_k + \frac{i}{M\omega_k} P_{-k} \right) \tag{3.7}$$

and

$$a_k^+ = A\left(Q_{-k} - \frac{i}{M\omega_k} P_k \right) \tag{3.8}$$

where, the quantity A is given as $\left(\dfrac{M\omega_k \pi}{h} \right)^{\frac{1}{2}}$.

Then the quantities a and a^+ are treated as quantum mechanical operators and are made to obey bosonic commutation properties. The entire lattice Hamiltonian can then be diagonalized in terms of these Bosonic operators (Abrikosov, Gorkov, and Dzyaloshinskii, 1963; Fetter and Walechka, 1971; Schrieffer, 1983). These Bosonic operators viz. a and a^+ correspond to "phonon" destruction and creation operators respectively. The energy eigenvalues associated with the phonon number operator, are the phonon energies. This is described mathematically as:

$$H_{lattice} = \sum_k \frac{h}{2\pi} \omega_k \left(a_k^+ a_k + \frac{1}{2} I \right) \tag{3.9}$$

where, I is the identity operator.

The phonons, as introduced above, are the quantum particles and obey Bose statistics with zero chemical potential. Moreover, they are spin zero particles. They in fact are the quasi-particles for collective lattice dynamics. They are non-interacting within the harmonic approximation for the dynamics and represent stable excitations. The above Hamiltonian also exhibits the existence of a zero-point energy contribution, a pure quantum effect having no classical analog. This is due to the uncertainty relation involving position coordinates and momenta of the lattice ions. This zero-point energy arises from the zero point vibrations of the lattice. These vibrations, when expanded in terms of the phonon operators, lead to the concept of zero-point phonons or virtual phonons.

In this context, it is advisable to briefly dwell on the Einstein model for phonons. These phonons are dispersionless, i.e., $\omega_k = \omega_E$.

A very important and interesting fact is that the Einstein model can be a realistic description for optical phonons (nearly dispersionless) (to be discussed below). Thus, the Debye model (appropriate for acoustic phonons) and Einstein model (suitable for optical phonons) in a combined way can lead to a good understanding for the thermodynamic and even optical properties of a crystalline solid.

The phonons are very important for both thermodynamic properties and transport properties of metals and semiconductors, as it is one of the most common elementary excitations. Besides, the nature of phonon dispersion is quite distinct for various types of condensed matter systems. The bare theoretically obtained phonon dispersion curve for harmonic phonons brings out the nature of the effective interatomic interaction W; whereas the experimentally observed phonon spectra throws light on the anharmonicity due to the interaction amongst various phonon modes themselves and also on the possible interaction between phonons and various other electronic excitations in the system. The experimentally obtained phonon spectra can be understood theoretically by carrying out full quantum many-body calculations involving various degrees of freedom in a solid. We will come back to this issue again later.

For dia-atomic lattice (having atoms of two different masses) the phonon spectra is rather complex. In that case, we will have 2 distinct equations for the different species with masses M_1 and M_2 corresponding to Eqn. (3.1). The equations of motion of the atoms/ions on the lattice, now involve two different ions and hence two different force constants. This causes the solutions for phonon spectra or the eigenmodes (phonons) to split into two distinct branches *viz.* acoustic branch and optical branch. The acoustic branch follows Born-von-Karmann-like dispersion; however, the optical branch has a finite frequency even in the long-wavelength limit and exhibits very weak dispersion (mathematical details are not presented here and can be found in the standard textbooks). It should also be pointed out that for a three-dimensional crystal, each of the above two branches has three longitudinal sub-branch modes and 6 transverse sub-branch modes. They are denoted by *LA*, *TA*, *LO*, and *TO*, where *L* denotes longitudinal, *T* denotes transverse, *A* stands for acoustic and *O* represents optical.

For condensed matter systems of experimental interest, all the above types of phonon modes themselves, through interaction amongst themselves or by coupling to conduction electrons can contribute to the lattice properties as well as electronic properties. The explicit lattice properties include lattice specific heat, lattice thermal conductivity, emergence of ferro- and anti-ferroelectricity, structural transition, etc. The electronic phenomena consist of transport processes involving polarons and also the formation of Cooper pairs for superconductivity, to name a few. These will be discussed at length in due course.

Another aspect of phononic Hamiltonian is the "anharmonicity," neglected in the bosonized version of the lattice Hamiltonian, as talked about earlier. The real solids exhibit anharmonicity and this is manifested in the thermal expansion. The origin of the anharmonicity lies in phonon-phonon interaction, neglected in the harmonic approximation used for quantization. With the inclusion of anharmonicity, the resulting Hamiltonian becomes very complex. This invariably leads to the softening (lowering of the frequencies of the phonon modes) and broadening (damping of the phonon modes) of phonon spectra. This has tremendous consequences for the phenomena of ferroelectricity/antiferroelectricity and structural transition seen in various solids, as well, and even leads to substantial enhancement in the magnitude of the attractive superconducting coupling constant in some cases, as we will see later in Chapter 5.

The melting of a solid is another phenomenon directly related to phonons. In fact, the melting temperature is intricately linked to the average number of phonons generated thermally. This is simply because when the mean vibrational amplitude of an atom becomes comparable to the interatomic spacing the solid melts, as was first proposed by Lindemann through a well-known criterion (Lifshits, 1982). Thus, the phonon energy spectra have a very important influence on the melting temperature of a solid.

3.2 ELECTRON-PHONON INTERACTION

The electron-phonon interaction arises very naturally in the elementary Hamiltonian used for describing the coupled electron-ion system for a crystalline solid. The deviation of the ion cores from the lattice site positions leads to a non-translationally invariant contribution to the effective potential experienced by a single electron due to the positive ion background. This contribution to the effective potential generated can be related to the ionic displacements. These displacements in turn can be expressed in terms of phonon operators and this leads to the emergence of an electron-phonon interaction term in the Hamiltonian. This can be sketched mathematically in the following way (Kittel, 1953):

$$U_{ei}(\mathbf{r}, \mathbf{R}_v^0) = U_{ei}(\mathbf{r} - \mathbf{R}_v^0) + \frac{\delta U_{ei}(\mathbf{r} - \mathbf{R}_v)}{\delta \mathbf{R}_v}\bigg|_{\mathbf{R}_v = \mathbf{R}_v^0} \delta \mathbf{R}_v \qquad (3.10)$$

where, $\mathbf{R}_v = \mathbf{R}_v^0 + \delta \mathbf{R}_v$. Furthermore, we carry out a summation of the potential U_{ei} occurring in the above equation over the lattice sites v and generate an effective potential $U_{eff}(\mathbf{r}_i)$ experienced by an electron with coordinate $\mathbf{r}_i$. From this $U_{eff}(r)$ we can perform Fourier transform to get U (q). Thus, the gradient of $U_{eff}(r)$ can easily be calculated. This leads to the electron-phonon interaction potential given by the following Hamiltonian:

$$H_{e-ph}(\mathbf{r}) = \frac{-i}{N} \sum_q \exp(i\mathbf{q} \cdot \mathbf{r}) U(\mathbf{q}) \mathbf{q} \cdot \left(\sum_v \delta \mathbf{R}_v \exp\left(-i\mathbf{q} \cdot \mathbf{R}_v^0\right) \right) \qquad (3.11)$$

where, we have used the property $\delta \mathbf{r}_i = -\delta \mathbf{R}_v$ and have retained $\delta \mathbf{R}_v$ in the expression (which is spatial Fourier Transform of Q_k), which can be expanded in terms of the phonon operators introduced earlier.

The electron-phonon interaction Hamiltonian takes the following form in the second quantization representation:

$$H_{e-ph} = \sum_{k,k',\lambda,\sigma} g_{k,k',\lambda} \Phi_{k-k',\lambda} C_{k',\sigma}^+ C_{k,\sigma} \qquad (3.12)$$

where, $g_{k,k',\lambda}$ is the matrix element of the electron-phonon interaction potential as given by Eqn. (3.10), between the Bloch states $|k\rangle$ and $|k'\rangle$ and λ is the polarization index of the phonon. The quantity Φ is given by:

$$\Phi_{q,\lambda} = \left(a_{q,\lambda} + a_{-q,\lambda}^+ \right) \qquad (3.13)$$

Thus, we can see from Eqn. (3.11) that the total electron-phonon interaction potential depends very strongly on the spatial gradient of the electron-ion potential and also on the phonon dispersion curves of all the phonon branches in the lattice. The electron-phonon interaction potential determines the coupling between the band electrons and the phonons at a spatial point r. In contrast, the electron-phonon interaction Hamiltonian given by Eqn. (3.12) describes the process of the scattering of a Bloch electron by the absorption or emission of a phonon.

The consequences of the electron-phonon interaction can primarily be classified into four types *viz.* (i) emergence of electrical resistance, (ii) formation of polaron, (iii) driving a Peierl's transition particularly in 1D systems, and (iv) mediating an effective electron-electron interaction. The first three effects arise from the first-order processes; whereas the fourth effect occurs when the second-order processes are considered. An electron prepared in a Bloch state with a particular quasi-momentum k under the influence of 1st order electron-phonon interaction gets scattered to another Bloch state with quasi-momentum k'. This is the origin of electrical resistance in a metal. This process takes place involving real or thermal phonons only. The other important process in the first order is that of the emergence of a new quasi-particle called polaron. It arises because the first order electron-phonon interaction can also cause a Bloch electron to be surrounded by a phonon cloud (mostly virtual phonons), while propagating. This also leads to an enhanced mass of this composite object.

This very composite object is known as "polaron." Its charge is however that of an electron. Mathematically a polaronic state, generated by 1st order perturbative correction due to electron-phonon interaction, is defined as:

$$|Polaron\rangle = |k,0\rangle + \sum_q |k,1q\rangle\langle k,1q| H_{e-ph} |k,0\rangle \frac{(-1)}{\omega_q} \qquad (3.14)$$

The polarons play very important roles in the transport properties of semiconductors and metals. The expectation value of the number of phonons surrounding a Bloch electron depends however upon the strength of the electron-phonon coupling. In the case of weak polaron, the above expectation value is much less than 1. In contrast, in the strong polaron limit, the number of accompanying phonons can even exceed 1. The energy of a polaron is generally calculated by the variational principle of quantum mechanics. This energy is often found to be less than the sum of the individual energies of the Bloch electron and the phonon. In that case, the polaron exists truly in a bound state of electron and phonon and can be considered to be a genuinely good Fermionic quasi-particle.

In the case of very strong coupling between the electrons and phonons, two polaronic quasi-particles can even bind further to form a new composite called "bipolaron," in a process very similar to the Heitler-London theory of molecule formation. Bipolaron is a Bosonic particle with charge $2e$ but mass substantially different from $2m_e$ and also from $2m_{pol}$.

In the case of 1D (or rather quasi-1D to be realistic) metallic or conducting systems where the Fermi surface shows nesting property, the 1st order electron-phonon interaction connects the states in the upper and lower subband and causes a dimerized lattice with an insulating gap to be formed and stabilized. This type of metal to insulator transition is known as Peierl's transition, as was first shown by R. Peierls. For higher-dimensional lattices like 2D (quasi-2D to be more precise) or 3D, it has been shown that if the Fermi surface has some parts which satisfy the nesting property, then Peierl's transition can occur (Ginzburg and Kirzhnits, 1982). This transition is often accompanied by the formation of a charge density wave (CDW) with the contribution from both lattice ions and the band electrons.

The most dramatic effect occurs when the second-order electron-phonon interaction is considered. This mediates an effective electron-electron interaction after the phonon degrees of freedom are averaged out. This can be shown by applying Schrieffer-Wolf transformation (SWT) (Kittel, 1963; Lifshits, 1982) to the Hamiltonian consisting of the contributions from non-interacting electrons, non-interacting phonons, and the first order electron-phonon interaction. Applying this standard scheme based on canonical transformation, one eliminates the first order electron-phonon interaction (Kittel, 1963). Then the lowest order coupling between the electrons and the phonons appears in the second-order only. Thus, the process now involves only virtual phonons *viz.*, the phonons which are non-thermal and are created and subsequently destroyed. The effective interaction mediated between the electrons by this virtual phonon exchange can be calculated from this term. Detailed calculation by Frohlich et al. shows that this term can become attractive when the phonon energy is greater than the change in energy of each of the participating (scattering) electrons near the Fermi surface (Kittel, 1963). Thus, this provides one of the most feasible mechanisms for superconductivity in metals and metallic alloys, as was shown by Bardeen, Cooper, and Schrieffer by means of an elaborate microscopic theory. It must however be pointed out that the real phonons (or thermal phonons) are also present and involved in the first order electron-phonon scattering processes in actual materials at non-zero temperatures even in the superconducting phase. Besides, the phonon mediated attraction is retarded, i.e., time-delayed. The contributions of these processes and aspects can be taken care of in detailed many-body treatments within the microscopic theory for superconductivity, like Eliashberg theory. We shall discuss these issues in detail in Chapter 5 (Superconductivity).

One of the spectacular consequences of phonon softening, in the vicinity of $q = 0$ and $q = \pi/a$, is the emergence of the phenomena of ferroelectricity and anti-ferroelectricity respectively. It occurs mostly in insulators containing Titanates like Barium Titanates. The natural softening of phonons near a structural transition, due to the anharmonicity, causes a distortion of the unit cells containing positive and negative ions. More precisely, the relative displacement of the positive and the negative ions within a unit cell or the relative displacements of the unit cells themselves with respect to each other can occur, leading to ferroelectric or anti-ferroelectric ordering respectively. This process results in the emergence of spontaneous electrical polarization in the system. The temperature at which the system orders, the static dielectric constant of the material exhibits a peak. This is expected as the application of a small electric field produces a large electrical polarization in the system at the ordering temperature.

Approaching a structural transition or a transition from the para-electric phase to ferro(anti-ferro)-electric phase by a system containing itinerant electrons, can lead to a lot of interesting consequences as a result of the coupling between the soft phonons and the itinerant electrons. Quite often, this coupling turns out to be quite strong and may even lead to attractive interaction of large magnitude between the electrons. This sometimes brings into play a very interesting competition involving superconductivity, ferro(anti-ferro)electricity, and structural transition. Furthermore, the nature of superconductivity here can also be of two distinct types *viz.* (i) fermion pairing (Cooper pair formation) based or (ii) Bose condensation (condensation of bipolarons) based, depending upon the strength of electron-phonon coupling to a large extent. This issue will be taken up in greater detail in Chapter 5 discussing superconductivity.

Besides, theoretical calculations involving many-body techniques show that phonons are very strongly renormalized (softened) in itinerant (band) magnets (Kim, 1978). Thus even in magnetic superconductors exhibiting interplay between itinerant magnetism and superconductivity, phonons may sometimes play an important role besides spin fluctuations, to mediate pairing interaction.

The experimental study of phonon spectra, in particular the dispersion relation, is carried out mostly with Raman or Brillouin scattering and with inelastic neutron scattering. The Raman spectroscopy can analyze the optical phonons; whereas the Brillouin scattering is used to detect the acoustic phonons. The inelastic neutron scattering technique, on the other hand, can probe both kinds of phonons in condensed matter systems. The first two techniques involve quantum interaction between the electromagnetic wave (more precisely the photons) and the phonons. The last technique is based on the interaction between neutrons and the nuclei of the ion cores situated at the lattice sites (Atland and Simons, 2006).

The techniques of Raman and Brillouin scattering are based on the inelastic scattering of photons from a medium, involving the absorption or emission of phonons. This leads to the well known Stokes and Anti-Stokes lines, corresponding to the enhancement or reduction of the incident photon frequencies. The theoretical formalism is based on the calculation of the frequency-dependent electronic polarization or dielectric function. The inelastic neutron scattering is based on the observed differential scattering cross-section of the neutron beam from a condensed matter system due to the lattice vibrations. The interaction of the incident neutrons with nuclei of the ion cores situated on the lattice sites is modified because of the lattice dynamics. The application of Born scattering theory leads to the proportionality between the differential scattering cross-section and the dynamical structure factor of the phonons (Lovesey, 1984). From the form of the dynamical structure factor (Fourier transform in wave-vector and energy space of the Spatio-temporal correlations involving the ion core displacements), one can extract the information regarding the phonon spectra including its dispersion relation. Thus, the phonon softening or even hardening in some situations can be directly observed through the above spectroscopic techniques.

Last but not least, the phonon spectra in spatially disordered solids remain another challenging problem. This assumes very great significance in view of the fact that a large number of novel phenomena discovered recently in condensed matter physics like quantum Hall effect (QHE) and high-temperature superconductivity, occur in disordered materials. The extensive research both theoretical and experimental has brought out that in the disordered systems, the phonons are fairly well defined in the long-wavelength regime. For large wave-vectors however, the departure from the translational symmetry becomes more prominent and effective. As a result, the phonons are heavily damped and become ill-defined (Wang et al., 2015).

3.3 ELECTRON-PHOTON/PHOTON-SOLID INTERACTION

The interaction between electrons and photons are very important in understanding various properties of a condensed matter system. In this section, we will mostly discuss those interaction processes in which electrons are excited into higher one-electron states within the band model by photon absorption. To start with, we assume for simplicity that the initial state belongs to the valence band and the final state lies in the conduction band. This process, therefore, describes optical absorption in semiconductors and insulators.

The interaction of photons with a solid may sometimes turn out to be very exciting, in the case of strong coupling between the photons and other elementary excitations of a solid. This coupling also arises from the electromagnetic interactions only. The excitations referred above are optical phonons and excitons (bound particle-hole pairs) in particular. In this case, of strong coupling, this interaction can no longer be treated by perturbation theory. Photon-phonon or photon-exciton is then intricately linked and the concept of a new quasi-particle emerges in a natural way. This novel elementary excitation is known as "polariton." It may be viewed as a photon surrounded by a cloud of phonons or excitons. They are known as "phonon-polariton" and "exciton-polariton" respectively (Madelung, 1996). They arise when we diagonalize a Hamiltonian consisting of photon field term, polarization (electronic or ionic indicating excitonic or phononic, respectively) field term, and a rather strong coupling term describing the interaction between photon field and quantized polarization field. Mathematically it can be described as:

$$H = \sum_k \left[\in_{1k} \left(A_k^+ A_k + \frac{1}{2} \right) + \in_{2k} \left(B_k^+ B_k + \frac{1}{2} \right) + \in_{3k} \left(A_k^+ B_k - A_k B_k^+ - A_k B_{-k} + A_{-k}^+ B_k^+ \right) \right]$$

(3.15)

where, A and B represent the photon field operator and the quantized polarization field operator respectively with ε_{1k} and ε_{2k} as the corresponding wave-vector dependent energies (Madelung, 1996). As explained before, the quantized polarization field can either be phononic or excitonic. Thus, the first term in the above Hamiltonian is the contribution from the photon field, the second term is that from the quantized polarization field, and the third term describes the interaction or coupling between these two fields. The signs of various terms in the interaction part have been fixed according to the product of the two fields E (electric) and P (polarization).

It is possible to diagonalize the above Hamiltonian by introducing "polariton operators" which are constructed as the linear combination of photon and polarization operators. The energy eigenvalues of this Hamiltonian give the dispersion corresponding to the various branches of excitation, as obtained in this scheme. Detailed analysis shows that one of the branches is similar to the photons in its dispersion and another one similar to the polarization quanta, with significant deviations appearing at low values of k.

One of the most important quantities relevant to the study of optical absorption is a real part of optical conductivity or imaginary part of dynamic dielectric function in the optical frequency range. They can be related using the standard Maxwell's equations and also constitutive equations. The optical absorption is directly proportional to the above-mentioned quantities. The shape of the curve describing the real part of optical conductivity $\sigma(\omega)$ versus ω gives us a lot of information regarding the nature of the excitations in the condensed matter system under consideration. For a metallic system, as we will see later, it can bring out the importance of many-body effects inherent in the system. This becomes particularly very useful when the metallic phase deviates from the conventional one.

Another important property of utmost importance in the electron-photon interaction process for condensed matter systems, with a special interest for the understanding of various types of conducting phases, is photo-emission (or inverse photo-emission) intensity as a function of photon energy. In this process, the beam of incident photons knocks out electrons from the system, in the case of the photo-emission process. In the other process *viz.* that of inverse photo-emission, the electron beam incident on a condensed matter system causes the emission of photons. Again, these experiments can throw a lot of light on the role of electron correlations in the condensed matter systems by extracting microscopic parameters like electron self-energies (both real and imaginary parts), electronic density of states (EDOS), etc.

Let us first carry out a fairly detailed analysis of optical absorption. This process may broadly be divided into two types *viz.* direct transitions and indirect transitions. In the case of direct transitions, the change in the wave vector of the electron is zero, i.e., $q = 0$. In the case of indirect ones, however, we have $q \neq 0$. Thus, the first process is purely a photonic absorption; whereas the second one is accompanied by an absorption or emission of phonons. Various classes of semi-conductors belonging to both direct and indirect types, exhibit these distinct phenomena (Kittel, 1953).

The absorption coefficient is defined as the energy flux absorbed per unit time divided by the incident energy flux per unit time. If $N(\omega)$ is the number of photons absorbed per unit volume per unit time, then $\dfrac{h}{2\pi}\omega N(...)$ is the energy absorbed per unit volume and unit time.

In the case of direct transitions, our detailed calculation relates the imaginary part of the dynamic dielectric function with the absorption flux in the following way:

$$K = \frac{(2h/2\pi)}{(\varepsilon_0 nc\Omega\, A_0{}^2)\, W(\Omega)} \tag{3.16}$$

where, ε_0 is the electric permittivity of the vacuum, A_0 is the amplitude of the space time-dependent vector potential, K is the absorption coefficient, $W(\omega)$ is directly related to the imaginary part of the complex dielectric function and ω is the frequency of the optical wave.

The function $W(\omega)$ can be calculated in a quantum mechanical procedure very similar to the Fermi Golden Rule (by taking into account transitions between various states) and is given by:

$$W(\omega) = \sum_{jj'} \frac{2}{t(2\pi)^3} \int W(j,j',\mathrm{k},\omega,t) d\tau_k$$

(3.17)

where, the function in the integrand in the above equation can be evaluated with the coupling between the photon field and electrons taken in the standard form (assuming Coulomb gauge) as:

$$H = \frac{e}{m} \mathrm{p} \cdot \mathrm{A}$$

(3.18)

Direct transition can be used to determine the shape of the absorption edge in semiconductors, provided the valence band maximum and the conduction band minimum lie at the same wave-vector.

In the case of indirect transition, however, the perturbation part consists of two contributions. The first part is the electron-photon coupling described earlier. The second part is the electron-phonon coupling term, which arises specially in this case.

There are however several possibilities within this process. We assume here that the principal contribution to the transition energy is carried by the photon and the main contribution to the transition momentum is taken by the phonon. Furthermore, we only consider contributions from LA-phonons. Again, the entire transition is regarded as a two-step process. In one step, the state of the electron changes by the absorption of a photon, in the other step it changes by the absorption or emission of a phonon. The intermediate state is a virtual state. It can be shown that energy is only conserved at the end of the entire process. Momentum although will have to be conserved in each of the individual steps, to keep the transition matrix element non zero.

In the calculational scheme for this case, the electron-photon coupling is taken as before. The electron-phonon coupling is the new contribution which is modeled as the first-order coupling between the band electrons and the LA-phonons. Again performing a calculation along the line of Golden Rule, leads to an expression relating the imaginary part of the dielectric function to the absorption and emission rates.

There are also other important and interesting processes like two-photon absorption. In the first step, an electron gets excited into a virtual state by the absorption of a photon of frequency ω_1 and polarization e_1. In the next step, it makes a transition to the final state with the absorption of a second photon of frequency ω_2 and polarization e_2. The probability of occurrences of such processes is very small. Therefore, to observe them experimentally, light sources of high intensity like lasers are needed.

Besides, there are processes involving the absorption of a photon and emission of a photon with a different energy, accompanied by the emission of one or more phonons.

When the phonons involved in these processes belong to the optical branch, the phenomena are known as "Raman Scattering." If they belong to the acoustic branch, the effect is called "Brillouin Scattering." These processes had been discussed earlier briefly in the context of phonons as well.

Another quite novel absorption phenomenon occurs involving excitons, the bound electron-hole pairs. When the photon energy is in the neighborhood of the absorption edge, the excitonic correlations play very important roles in determining the absorption spectrum of a solid. This affects both direct and indirect transitions and exciton creation is assumed to be a true absorptive mechanism. Here again, the imaginary part of the dielectric constant can be related to the various transition matrix elements for both allowed and forbidden transitions. These expressions also involve the binding energy of the exciton. The imaginary part of the dielectric constant shows an appreciable enhancement at energy near the absorption edge. Even at energies much above the absorption edge, the exciton effects are displayed in the spectrum for the imaginary part. The experimental absorption spectra obtained from various semiconductors, particularly in the vicinity of the absorption edge, agree quite well with the above theoretical predictions involving the phonon and exciton mechanisms (See for example Madelung (2004), Chapter 6, p. 284, Figure 6.9).

In addition to the transitions between different bands, as we have considered so far, there can be optical transitions between the electronic states within the same band as well. These processes are quite important when the occupied (initial) and unoccupied (final) states are present within one band. This is realized in metals and some of the semiconductors. This situation can be treated as an interaction between electromagnetic waves in the optical region and gas of quasi-free electrons, characterized by an effective band mass, in the vicinity of the band edges. The electron-phonon interaction will have to be involved to achieve a transition between different $|k\rangle$ states of a band electron. Then the absorption of light by the quasi-free charge carriers can be treated as a transport problem. The standard classical equation of motion in the relaxation time approximation leads to expressions for real and the imaginary part of the dynamic dielectric function. The dc conductivity obeys Ohm's law, as expected. The longitudinal dielectric function also exhibits the electron plasma collective modes. The absorption behavior is decided by the position of the frequency of the electromagnetic wave with respect to the plasma edge, to a large extent. To elaborate, for frequency less than the plasma frequency the optical wave gets reflected from a metal. However, for light with a frequency above this plasma edge, the system again becomes transparent. These results are borne out by the analytical expressions of the real and imaginary parts of the dielectric function for quasi-free band electrons within the RPA. The detailed absorption and reflection coefficients in the presence of electron-phonon coupling can also be worked out. The imaginary part of the dielectric function in the coupled electron-phonon system decides this. The electron-exciton coupling also leads to the formation of many other quasi-particles like 'exciton polaron' and 'transferon' in magneto-excitonic semiconductors (Nagaev, 1983).

The remaining important issue is the quantum interaction of light directly with lattice vibrations. The process involves photons and TO-phonons and is seen in polar solids. The simplest processes are the transformation of a photon into a phonon of the same energy and wave vector. The next (second-order processes) are events in which a photon decays into two phonons or into a photon and a phonon. Furthermore, lattice anharmonicity can lead to an additional process in which one phonon decays into two phonons. These mixing processes of photons and phonons lead to the emergence of new quasi-particles named "polariton," as described before. The possibility of the presence of multi-phonon processes makes the polaritons even more exciting. The electron-hole (excitonic) pairs may also be involved in the absorption of photons and lead to different types of polaritons, as stated before. The properties of the polaritons (both phonon-polariton and exciton-polariton) and in particular, their dispersion relations, are very important for understanding the optical absorption spectra of metals and semiconductors, especially in the form of quantum wells and even nanowires (Agarwal and Jha, 1982; Wang et al., 2011, 2014; Balaguru and Jeyaprakash; Shi et al., 2016). The existence of the composite entity comprising of the photon and polarization quantum (phononic or excitonic origin) as a new bosonic excitation, has lots of important consequences. The two emerging branches of this new quasi-particle possessing different dispersions carry the signature of the various microscopic processes taking place in the solid (see Madelung (2004), Chapter 6, Figure 6.3).

Another exotic effect involving electrons and photons, called 'photo-magnetism' or more precisely 'photo-ferromagnetism' was proposed theoretically and has been partially verified experimentally in the magnetic semiconductors like the rare earth chalcogenides (Nagaev, 1983). The central idea is that a sufficient number of photoexcited carriers (electrons) in the conduction band can induce indirect exchange amongst the ordered spins of the localized type and are thus expected to raise the Curie temperature of a ferromagnet and may even be able to convert an antiferromagnet into a ferromagnet (Berdyshev, 1962; Karpenko and Berdyshev, 1963). There are also evidences of spin-excitons in heavy-fermion semi-metals (Riseborough and Magalhaes, 2016). Besides, 'photo induced magnetism' in semiconductors and various 'optoelectronic processes' related to semiconductor-based devices in particular at the mesoscopic and nanoscales, have become very important in the present-day research.

KEYWORDS

- **antiferromagnet**
- **chalcogenides**
- **charge density wave**
- **Curie temperature**
- **ferromagnet**
- **photo-ferromagnetism**

REFERENCES

Abrikosov, A. A., Gorkov, L. P., & Dzyaloshinskii, I. E., (1963). *Methods of Quantum Field Theory in Statistical Physics.* Dover Publications, Inc. New York.

Agarwal G. S., & Jha, S. S., (1982). *Solid State Communications, 41*(6), 499.

Ashcroft, N. W., & Mermin, N. D., (2017). *Solid State Physics.* Brooks Cole/ Cengage Learning India.

Atland, A., & Simons, B., (2006). *Condensed Matter Field Theory.* Chapter 5, Cambridge University Press, Cambridge.

Balaguru, R. J. B., & Jeyaprakash, B. G.*NPTEL-Electrical and Electronics Engineering: Semiconductor Nanodevices.*

Bartlett, R. J., (1997). Recent advances in coupled-cluster methods. In: *Recent Advances in Computational Chemistry* (Vol. 3). World Scientific.

Berdyshev, A. A., (1962). *FTT, 4,* 1382.

Cox, P. A., (2010). *Transition Metal Oxides: An Introduction to Their Electronic Structure and Properties.* Oxford.

Fetter, A. L., & Walecka, J. D., (1971). *Quantum Theory of Many Particle Systems.* McGraw-Hill, San Francisco.

Ginzburg V. L., & Kirzhnits, D. A., (1982). *High Temperature SuperconductivityPhys. Rep. C4, 344.* (New York: Consultants Bureau).

Huang, K., (1975). *Statistical Mechanics.* Wiley Eastern Private Limited, New Delhi.

Karpenko, B. V., & Berdyshev, A. A., (1963). *FTT, 5,* 3397.

Kim, D. J. (1978) . *Phys. Rev. B 17,* 468

Kim, D. J. (1980). *Solid State Commun. 34,* 963.

Kittel, C., (1953). *Introduction to Solid State Physics.* Wiley, New York.

Kittel, C., (1963). *Quantum Theory of Solids.* Wiley, New York.

Lifshits, I. M., (1982). *Quantum Theory of Solids.* Mir Publishers, Moscow.

Madelung, O., (1996). *Introduction to Solid State Theory.* Springer-Verlag, Heidelberg.

Mahan, G. D., (1981). *Many Particle Physics.* Plenum Press.

Nagaev, E. L., (1983). *Physics of Magnetic Semiconductors.* Mir Publishers, Moscow.

Pines, D., (1999). *Elementary Excitations in Solids.* Perseus Books Publishing, Massachusetts.

Riseborough, P. S., & Magalhaes, S. C., (2016). *J. Magn. and Magn. Mat., 400.*

Schrieffer, J. R., (1983). *Theory of Superconductivity.* Advanced Books Classics, Dover.

Shi, L., Wang, C., Wang, J., Fang, Z., & Xing, H. *Advances in Materials Physics and Chemistry, 06*(12), 305. doi: 10.4236/*ampc.*2016.612029.

Wang, J., Chen, Z., Liu, Y., Shek, C. H., Lawrence, C. M., & Wu, L. J. K. L., (2014). *Solar Energy Materials and Solar Cells, 128,* 254.

Wang, K., Chen, J., Rai, S. C., & Zhou, W., (2011). In: Zhou, W., & Wang, Z. L., (eds.), *Chapter 16; Three Dimensional Nanoarchitectures: Designing Next-Generation Devices.* Springer, New York.

Wang, X., Zheng, W., Wang, L., & Xu, N., (2015). *Phys. Rev. Lett., 114,* 035502.

CHAPTER 4

Electrons in Solids and Surface States

4.1 INTRODUCTION

The problem of electrons in solids, both in the bulk and at the surface, is a very old problem. It is almost as old as the solid-state physics itself. Understanding of this requires quantum treatments. The study of interacting electrons on lattices is an important component in this endeavor. Interacting electron gas is a rather simple but non-trivial model for this in condensed matter physics. It is as important as the band theory of solids. The high electron density regime is relevant to the alkali metals; whereas most of the other metals fall in the intermediate density regime. On the other hand, the very low-density regime is very fascinating and leads to an insulating phase known as "Wigner Solid (WS)" (Pines, 1999).

The conventional perturbation theory in the framework of weakly interacting gas of electrons, with no band-structure effects included, has been applied to both high density and intermediate-density electron gas in three-dimension with a lot of success. The simple metals and metallic alloys are fairly well described by the above treatment. The transition metals and their compounds however are often found to show marked deviation from the predictions of the above treatment. This brings us to the possibility of carrying out the perturbation theory in the reverse order *viz.* treating the interaction part as a principal term and the kinetic energy part as the perturbation term. This is well realized in the situation with low-density electron gas in three-dimension. As a consequence of this unique perturbation theoretic calculation, the electrons become spatially localized. This leads to the formation of so-called "electron solid," also known as "WS" (as mentioned earlier), where electrons become immobile and the system becomes insulating. The strongly correlated electron system described by the conventional Mott-Hubbard model (although a lattice model but does not invoke the band structure effects in detail) in the very large Coulomb correlation limit, finds a shadow in Wigner's Solid scenario.

In the theoretical treatments described above corresponding to electron gas, the band structure effects have been completely ignored. The inclusion of the presence of real lattice makes the calculations extremely difficult and even intractable.

The interacting electron gas problem in a high-density limit can be handled by standard perturbation theory, leading to Hartree-Fock and Correlation corrections to the ground state energy. These calculations yield various physically important quantities such as cohesive energy, specific heat, magnetic susceptibility, and charge response like dielectric function amongst others. They are in quite good agreement with experimental results from simple alkali metals. For most of the other metals, barring the transition metals, the electron gas in the intermediate density regime turns out to be a more suitable model. The perturbation theoretic scheme, as used in the high-density limit, needs some non-trivial refinement in this case. These corrections are of many-body origin and involve delicate handling of the screened Coulomb repulsion in the realistic intermediate electron density regime.

The standard perturbation theory aims to calculate the ground state energy of the three-dimensional interacting electron system, as mentioned above. There are other schemes for applying this technique for calculating electronic polarizability and electronic Green function as well. These quantities play very important roles in describing the microscopic quantum state and the nature of the interacting electron system.

The case of a strongly correlated electronic system as described by the WS or Mott-Hubbard model will also be taken up. Here the inter-electronic Coulomb interaction plays the more dominant role. The kinetic energy term being of much smaller magnitude compared to the Coulomb repulsion can be treated as a perturbation. Increasing the kinetic energy gradually leads to the "melting" of WS or analogously brings a Mott-Hubbard insulator (MHI) to a conducting state under certain conditions. It is believed that this melted WS or conducting MHI is characterized by a quantum state which may be very different from that of a conventional metallic system.

We will describe the above features and consequences of the interacting electron system in detail in this chapter. We will also present the various phenomenologies corresponding to the various possible states of the electron system. The most well-known amongst them is Landau's Fermi liquid theory, Tomonaga-Luttinger liquid (TLL) theory, and marginal Fermi liquid (MFL) theory proposed by Varma and co-workers. The weakly interacting three-dimensional electron gas both in the high and in the metallic density regime conforms to Landau's Fermi Liquid (LFL) phenomenology. The low-dimensional and strongly correlated electron systems are often characterized by the other two effective theories *viz.* TLL theory and MFL theory.

In the LFL picture, the low energy single-electron excitations from the Fermi surface in an interacting electron gas system in 3d, are put in one to one correspondence with those from the non-interacting system, i.e., ideal fermi gas. This leads to an effective mass of the excitations enhanced from the bare mass of an electron. This single electron excitation is known as a "quasi-particle." Accordingly, all the charge and spin response functions in the interacting case get related to their non-interacting counterparts in a rather simple way, involving the effective mass and some other interaction parameters known as Landau parameters.

In the TLL scenario, the Fermi surface is non-existent in the true sense and the spin degree of freedom and the charge degree of freedom of the single electron separate out or decouple. There are two kinds of quantized collective excitations represented by *viz.* spinless charged boson (holon) and charge-less spin 1 boson (spinon) which emerges as the new quasi-particles in this case. Each physical carrier *viz.* electron (or hole) splits into a holon and a spinon. This model was originally proposed for interacting electron systems in 1d. The caution must be applied however in its application to real condensed matter systems, as even the low-dimensional ones can at best be quasi-one-dimensional.

The MFL phenomenology retains the concept of a Fermi surface and also assumes that the spin and charge of carriers (electrons or holes) are intact, similar to as in LFL. It however brings out a novel concept that the quasi-particles are ill-defined exactly at the Fermi surface. Slightly away from this surface, however, they possess good spectral weight and stability. This model was put forward with a view to addressing the anomalous normal state properties of the strongly correlated systems observed in the under-doped to optimally doped stoichiometric regimes of the layered Cuprates.

The last topic we will briefly cover in this chapter discusses the effects arising from the finite size of a real solid. They include the elementary ideas regarding the emergence of 'surface states' and the occurrences of the phenomena of 'surface melting' and 'surface reconstruction.' As was pointed out briefly in Chapter 1, the lattice periodicity is broken at the surface of any finite-sized real solid, leading to a breakdown of Bloch's theorem. This causes new electronic states to be created in the forbidden gap region. These states describe spatially localized electronic wave functions, in contrast to the propagating Bloch states (band states) which are found in the bulk of a solid. There are also various effects connected with the destruction and reconstruction of a surface, arising from the lattice dynamics at finite temperature.

4.2   WEAKLY INTERACTING ELECTRON GAS

In this section, we take up the problem of the three-dimensional electron gas and describe the results of the perturbative scheme in some details. The model we use is known as the Jellium model. It comprises electron gas interacting via two-body Coulombic repulsion and a smeared out positive ion jelly which interact amongst itself and also with the electrons, again via Coulomb interaction of repulsive and attractive nature respectively. The positive ion jelly ensures that the charge neutrality is maintained locally everywhere and the system is in equilibrium.

The simplification of a discrete lattice to a continuum jelly model implies that the Bloch wave functions reduce to plane waves. The relevant electronic Hamiltonian becomes:

$$H^{el}_{jellium} = H_{kinetic} + H_{Coulomb} \tag{4.1}$$

where, the first term in the right-hand side is the total kinetic energy operator of the electrons and the second term describes the Coulomb interaction operator involving the electrons. Mathematically:

$$H_{kinetic} = \sum_{k,\sigma} \in_k C^+_{k\sigma} C_{k\sigma} \tag{4.2}$$

and

$$H_{Coulomb} = \frac{1}{2} \sum_{k_1 k_2 q \sigma, \sigma'} V_q C^+_{k_1+q\sigma} C^+_{k_2+q\sigma'} C_{k_2\sigma'} C_{k_1\sigma} \tag{4.3}$$

where,

$$\epsilon_k = \frac{h^2 k^2}{8m\pi^2}; V_q = \frac{4\pi e^2}{q^2} \tag{4.4}$$

We first carry out the first-order perturbation theory. The unperturbed ground state is the non-interacting Fermi Sea. It is expressed as:

$$|0\rangle = \prod_{k,\sigma} C^+_{k,\sigma} |vac\rangle \tag{4.5}$$

The expectation value of $H_{kinetic}$ in the ground state can be shown to be equal to $\frac{2.21}{r_s^2}$, where, r_s is the effective inter-electron distance, varying as $\rho^{(-1/3)}$ with ρ being the electron density (Pines, 1999). This then is the unperturbed ground state energy. The lowest order, i.e., the first-order correction arising from the inter-electron Coulomb repulsion consists of two parts viz. Hartree and Fock. The Hartree contribution is the expectation value of the direct part of the Coulomb interaction in the Fermi Sea. On the other hand, the Fock contribution is the corresponding expectation value of the exchange part of the Coulomb interaction. It can be shown that the Hartree contribution exactly cancels the total contributions from the electron-ion and the ion-ion interaction contributions. The Hartree-Fock part gives a negative contribution due to the anti-commutation property of the fermion operators. The correction to the ground state energy due to this comes out as $-1/r_s$. Thus it is clear that for high electron density, i.e., low values of r_s, in particular $r_s \ll 1$, the kinetic energy contribution dominates over the interaction contribution from the first order. It is worthwhile pointing out that the exchange contribution is meaningful only when the two interacting electrons possess a parallel spin configuration and it assumes a negative magnitude (Pines, 1999).

The higher-order perturbative corrections from the Coulomb interaction involving the pair of electrons with parallel spins and anti-parallel spins generate further exchange contributions and "correlation" contributions, respectively. The latter also gives a negative contribution and implies an attraction between electrons and holes corresponding to opposite spins. The perturbative corrections can be summed up to infinite order and the summation converges; although there may be singularities in the contributions from different orders of perturbation individually. This result of the finite correction to the ground state energy of the weakly interacting electron gas is crucially dependent on the assumption of the high electron density limit. In this density regime, certain approximations known as random phase approximations (RPA) (briefly described in the second chapter) are used to evaluate the contributions from different orders. This approximation selects a class of terms in the calculation of the contribution from a particular order."

The above calculation leads to estimates for the various physical quantities of interest like: (i) ground state energy, (ii) cohesive energy, (iii) electrical polarizability, (iv) spin susceptibility, etc. Two salient features of these calculations may be summarized as (i) the emergence of appreciable weightage for the tendency to form spin density wave (SDW) and (ii) the possibility of screened Coulomb interaction turning "attractive" at short distances. Besides, one sees the usual Stoner enhancement for the uniform spin susceptibility.

The extension of the above perturbative calculations to the intermediate density regime requires the need to abandon various analytical approximations adopted and valid in the high-density regime. This leads to better predictions and understanding of the properties of the real metals. For example, the estimates for the cohesive energy are much closer to those of the real metals (Shastry et al , 1974).

The above theoretical treatments and analysis also lead to very important concepts of effective low energy theory for interacting electron systems. The principal idea is that the electron-hole excitations very close to the Fermi energy for the weakly interacting system and the non-interacting system can be put in a one to one correspondence. The correspondence works with a constraint that the volume of the Fermi surfaces in the two cases is the same (Luttinger's theorem). The single-particle excitations in the interacting case are called "quasi-particles" to distinguish them from their counterparts (particles) in the ideal Fermi gas case. The quasi-particles carry different masses than the particles which have bare mass. The perturbative calculation for the weakly interacting electron gas can be used to determine the effective mass of the quasi-particles.

The spin response and the charge response of the interacting electron gas can be calculated by the detailed perturbative technique, as mentioned before. In the non-interacting case or even in the interacting case retaining terms up to Hartree level (as is done in RPA), we observe Pauli susceptibility for a spin response. This gets modified to Stoner form for the interacting case with the inclusion of exchange-correlation corrections in the lowest order. Taking into account higher-order exchange and correlation contributions will however change the form of the spin response from Stoner's expression. This also has been worked out in detail in both high density and intermediate-density regime. The competition between ferromagnetic and anti-ferromagnetic tendencies also emerges from this analysis, because of the opposite character of exchange and correlation effects. The comparison of the theoretical results with those from the real metals is however not so good. This is attributed to the important role of band structure effects in itinerant magnetism, manifested through the "local field corrections" involving the Umklapp vectors for the inclusion of exchange and correlation effects.

The charge response at the RPA or Hartree level exhibits the well-known Thomas Fermi screening and also the plasma oscillations. This is brought out very clearly in the standard equation of motion approach or by the Feynman Diagram technique. The basic physics behind these effects are described below. Any test charge placed in the electron gas will create an electric field. This field causes the electrons and the positive ions to move in opposite directions. This creates an electrical polarization in the system. Moreover, the motion of the charges viz. both electrons and the positive ions lead to a "screening" of the test charge. After the screening is complete, the local charge neutrality is restored once again. It may happen however that, during the motion of the charges they may sometimes overshoot the test charge. This will set an oscillation of the positive ions and the electron gas then. This is the origin of the plasma oscillation in the electron gas. The dynamic longitudinal dielectric function calculated from the polarization exhibits this plasma mode. The static dielectric function in the long-wavelength limit, i.e., in the case of a slowly varying applied electric field from the external charge, shows the existence of a spatially decaying electric field as a result of the screening. The mathematical expression of this quantity is of the well-known Yukawa form and is called the Thomas-Fermi dielectric function.

For the general wave-vector and frequency-dependent dielectric function for the weakly interacting electron gas, the RPA treatment leads to Lindhard's expression for the same. The response function becomes complex in this case. The zero of the real part gives the plasma mode. It should be highlighted that the plasma oscillations are a collective mode, as it involves the temporal oscillation of the density operator corresponding to the entire electron gas.

The expressions of Lindhard functions in 3d corresponding to the real part and the imaginary part of the longitudinal dielectric function denoted by $\varepsilon(q, \omega)_R$ and $\varepsilon(q, \omega)_I$ respectively, are very long and can be found in any standard textbook (for example in the book "Elementary Excitations In Solids" by David Pines and also in the published papers (Lindhard, 1954; Stern, 1957; Ruvalds, 1987)). In this chapter, we bring out some of the salient features of these response functions.

The function ε_R is symmetric in ω. It contains frequency dependence in a logarithmic part as well as in a quadratic part. The function ε_I on the contrary takes 3 different forms depending upon the regime of ω. In the static limit, the only ε_R survives and in the long-wavelength regime, it reduces to the Thomas-Fermi form.

The standard quantization scheme can be applied to the plasma modes to generate the bosonic quasi-particles "plasmons." The electronic Hamiltonian can be diagonalized in terms of the plasmon operators. The energy vs. wave vector dispersion curve of the plasmons in the long-wavelength limit is flat and it shows an energy gap indicating that the plasmon modes are "massive." For finite q values, the dispersion function becomes quadratic in q.

Stretching the calculations beyond RPA, one can see the damping of plasma modes due to the non-linear coupling between these modes. Moreover, the kinetic energy of the electrons causes a decay of the collective plasmon modes into the single particle-hole excitations at higher magnitudes of q beyond a critical threshold value. This decay process is known as "Landau Damping" (Pines, 1999).

There are several interesting properties of the RPA dielectric function. One of them is "Kohn Singularity." This is exhibited by the static dielectric function at zero temperature. It is described as the singularity of the derivative of the q-dependent dielectric function with respect to q at $q = 2k_F$, where k_F is the Fermi wave-vector. This property is the consequence of the existence of the sharp Fermi surface or more explicitly the sharp discontinuity of the electron occupational distribution function, i.e., the Fermi distribution function at the Fermi wave-vector k_F for $T << T_F$. This can lead to another interesting phenomenon known as "Kohn Anomaly" manifested in the singularity of phonon dispersion function, when the electron-phonon coupling is taken into account. Recently it has been shown that the ideas of Kohn Singularity and Kohn Anomaly can be extended to even superconductors (Chaudhury and Das, 2010–2013). This would be discussed in greater detail in Chapter 5.

Another very important effect associated with the presence of the sharp Fermi surface is "Friedel oscillations." This is seen when an external charge is placed in a metal. The spatial density of the screening electrons of the metal exhibits an oscillatory dependence combined with power-law decay with distance from the external charge. The electrostatic potential due to the external charge also displays this same behavior. This phenomenon has various other manifestations. The notable amongst these is the well known"RKKY (Rudolf-Kittel-Kasua-Yoshida) interaction" experienced by the magnetic impurity atoms situated in a non-magnetic metal (Yamada, 2004). We have discussed this briefly in Chapter 2 on Magnetism. Both Friedel oscillations and RKKY interaction are prominent at zero temperature; however, they get exponentially damped at finite temperature because of the nature of the Fermi distribution function.

As mentioned before, the weakly interacting high-density electron gas can be put into one to one correspondence with the ideal electron gas, when the perturbation treatment is carried out to all orders. This is the microscopic basis for the phenomenological LFL, as was shown by Gellmann, Bruckener, Kohn, and Luttinger(Gell-Mann, 1957; Pines, 1999; Yamada, 2004). This was further extended to the case of intermediate electron densities appropriate to many of the real metals (Shastry, 1978; Shastry et al., 1974). Thus, the concept of the Fermi surface still remains intact even in most of the real metals. This enables us to observe experimentally the various effects connected with charge and spin responses, described above.

The basic starting point of the LFL theory is the postulate of the existence of quasiparticles in an interacting Fermi system. These quasi-particles are analogous to the bare particles in the non-interacting system, in that they play very similar roles. The quasi-particles near the Fermi surface are almost non-interacting and are therefore stable and have an effective mass. The very weak interaction amongst them is characterized by Landau parameters and is related to the functional derivatives of the total energy of the system with respect to the number of quasi-particles. The effective mass of these quasi-particles is related to the Landau parameters. The Landau parameters are of two types *viz* spin symmetric and spin anti-symmetric. The properties like the density of states and various charge response functions of an interacting electron system can be expressed in terms of the spin symmetric Landau parameter. The spin susceptibility and various other spin-dependent properties can be described with the help of spin anti-symmetric Landau parameter. The possible enhancement of the magnitude of the specific heat and modification of the spin susceptibility from the expected Pauli type of behavior in real metals known as Stoner enhancement can be understood satisfactorily with the help of these phenomenological Landau parameters. In most of the common metals and even for some of the novel systems like Heavy Fermion Compounds in a large temperature regime, LFL provides a very useful description of the electronic state and a sound starting ground for the theoretical study of various other phenomena.

It must be emphasized, however, that unlike the continuum situation the extension or application of LFL theory to the case of lattices is not at all straightforward or rigorous.

Before we go over to the situation with strong electronic correlations, we take a look at the case of pure 2d interacting electron gas interacting via bare Coulomb repulsion of logarithmic nature, even in the high-density limit and its possible consequences to the case of layered systems (Chaudhury and Gangopadhyay, 1995). It was shown there that the low energy effective theory obtained by momentum space scaling, starting with the screened Coulomb interaction under RPA in this case, leads to a spontaneous breakdown of the LFL picture. Furthermore, with the effective screened interaction turning out to be attractive, it leads to the possibility of the formation of bound pairs of fermions in a natural way. For a review of the RPA results for the polarizability of the 2d electron gas including the plasma dispersion relation, the reader may look at the above reference and other papers (Ruvalds, 1987). Again, systems in the metallic electronic density regime with large electronic correlation contribution but within the 3d LFL scenario, can also exhibit very interesting microscopic physics leading to effective inter-electron attraction from the approach to magnetic ordering of itinerant nature (Chaudhury and Jha, 1984). This has been extended to the layered systems too and even with some multi-band effects (Chaudhury, 2009; Chaudhury and Das, 2018).

4.3 STRONGLY CORRELATED ELECTRONIC SYSTEMS

There are, however, systems where the conventional electron gas pertur-bation theory and the associated LFL phenomenology break down. These systems are known as "strongly correlated systems." As mentioned before, there are two types of phenomenologies proposed for these types of systems, *viz* TLL theory and MFL theory. The first and foremost feature of these systems is the failure of the complete description of the electronic state in terms of an effective single-particle picture as used in band theory. Therefore, the electronic models relevant to these phenomenologies necessarily involve discrete lattices. Moreover, the microscopic treatment will also have to be carried out in a very different fashion.

It should be also pointed out that the analogous system (for MHI) corresponding to the strongly interacting electron gas, i.e., WS also describes a frozen electron system on a discrete lattice in real space.

The microscopic Hamiltonians corresponding to the systems conforming to the above two phenomenologies must be in the regime of large U in the Hubbard model like the description. In one dimension the simple one band Mott-Hubbard model or simpler version of this model with $U >> t$, has been shown to behave as a TLL, when the number of electrons per lattice site is less than 1 (Tomonaga, 1950; Luttinger, 1963; Mahan, 1981; Atland and Simons, 2006). Moreover, the entire electronic system gets decomposed into two parts viz a subsystem of spin-less bosons and another of charge-less fermions. The distribution function of the carriers, either mobile vacancies (holes) or electrons deviates very strongly from that of Fermi distribution function. In particular, the discontinuity in the k-space occupation number for the carriers, at the Fermi wave-vector (k_F) vanishes in a power-law fashion. In higher dimensional lattices, there are no exact solutions of these microscopic models. Therefore, the scenario there is unclear. There are however conjectures regarding the emergence of TLL-like behavior on the two-dimensional lattice as well. The numerical calculations performed so far are not very conclusive; although they seem to indicate an anomaly around $k = k_F$. For a certain class of quasi-1D systems, however, including the organic superconductors, the TLL gives a good description of the normal state phenomenology at high temperatures.

Another slightly different model which is believed to lead to MFL phenomenology is the extended Hubbard model involving two (or more) bands. In this model besides U's, there is one more parameter V representing the Coulomb repulsion between the electrons on the atoms situated on the nearest neighbor sites. The introduction of V ensures stability for the system with ionicity effects present as well, besides the usual covalency.

The analysis of the multi-band extended Hubbard model is much more complex; however, it is a more realistic model for most of the strongly correlated systems belonging to transition metal oxides, having both covalency and ionicity effects. The Hamiltonian consists of both on-site electron-electron repulsion and the nearest neighbor electron-electron repulsive interaction term. The on-site term is taken to be very large compared to the bandwidth; whereas the second term is finite but may not be very large. The consequences of this model have been explored numerically for two-dimensional lattice. The situation with the number of electrons per site less than 1 is very fascinating. The carriers seem to develop an "effective attraction" amongst themselves and form bound states spontaneously under certain conditions in this case. This has several consequences. One of them is the marked deviation from the usual LFL behavior. There are several possible descriptions of the scenario in this case. Amongst these, a very promising candidate is MFL.

The MFL phenomenology is characterized by the following behavior of the single-particle irreducible self-energy function:

$$\Sigma_R(q,\omega) = \omega \ln \frac{|\omega|}{\omega_c}$$

(4.6)

and

$$\Sigma_I(q,\omega) = |\omega|$$

(4.7)

where, the indices R and I denote the real and the imaginary parts respectively of the self-energy Σ and ω_c is a cutoff frequency scale in the theory. The above relation holds at very low temperature viz $|\omega| >> T$. In the usual FL scenario in three-dimensions, the real, and the imaginary part goes as ω and ω^2, respectively. The peculiarities of the functional form of the self-energy in MFL lead to various interesting consequences. The most spectacular amongst them is the absence of a true quasi-particle at the Fermi surface. This is exhibited by the disappearance of the discontinuity in the single fermion occupation number n_k at k_F at zero temperature.

The above phenomenologies viz. TLL and MFL have been tried out for analyzing the experimentally obtained normal state properties for the layered Cuprate superconductors in the so-called "underdoped regime" where the effective inter-electron Coulombic repulsive interaction is believed to be much stronger in comparison to the bandwidth. The results of the analysis are however rather inconclusive; although there are overwhelming evidences in favor of MFL scenario. In the very low doping regime, however, there is a possibility of occurrence of Mott-Hubbard type of anomalous conductor, distinct from that with MFL character.

Interestingly, in the over-doped regions of the cuprates in the normal phase, the standard LFL phenomenology once again provides a very good description (Roy Chowdhury and Chaudhury, 2016, 2018).

Unlike LFL and TLL phenomenologies, however, the microscopic foundation for MFL scenario is still not firmly established (Chaudhury and Carbotte, 1993; Chaudhury, 1995); although there have been attempts to invoke 2d physics with the analogies from dissipative XY model lately in this direction (Varma, 2015).

Many aspects of TLL, like spin-charge decoupling and its possible relevance and application to the normal state description of the quasi-1D organic superconductors at high temperatures, have also been analyzed and discussed (Roy Chowdhury and Chaudhury, 2015). The feasibility of MFL-like scenario in lightly overdoped quasi-2D cuprates have also been investigated lately (Roy Chowdhury and Chaudhury, 2018). The microscopics of doped strongly correlated quantum Heisenberg antiferromagnets in both 2d and 1d have been investigated from the theoretical calculations for generalized stiffness constants with results for effective couplings for spin and charge degrees of freedom in the normal phase (Bhattacharjee and Chaudhury, 2016; Bhattacharjee, 2016; Bhattacharjee and Chaudhury, 2018). The results describe the evolution of magnetism from localized type to that of itinerant nature very clearly with emergence of new modes or phase separation and even change in the sign of the effective exchange coupling, depending upon the dimensionality of the lattice. This is very crucial for the microscopic understanding of pairing and the role of spin and charge fluctuations in the underdoped phases of cuprates inside the layer and also along the chain (Figure 4.1).

4.4 SURFACE STATES, SURFACE MELTING, AND SURFACE RECONSTRUCTION

A surface represents an abrupt transition between the perfectly periodic lattice and the vacuum. A surface itself is a two-dimensional periodic arrangement of atoms. This structure is either the same as the structure inside the solid or it is a superstructure created by the rearrangement of the surface atoms. In the former case, the surface is called an "ideal surface." If the surface is covered with an adsorbed layer, this also can have a structure, which deviates from the lattice structure inside the solid.

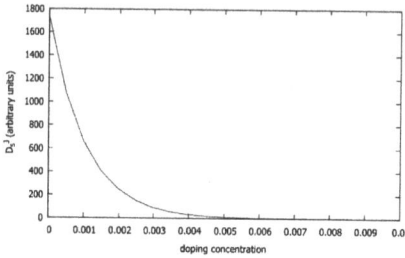

Figure 1a:

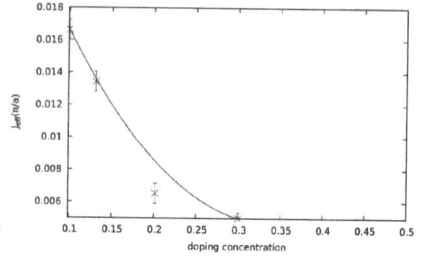

Figure 1b:

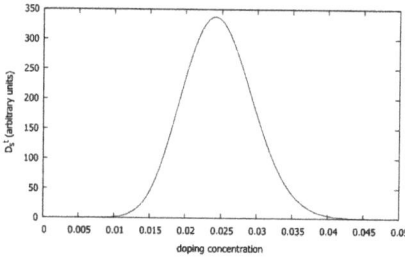

Figure 1c:

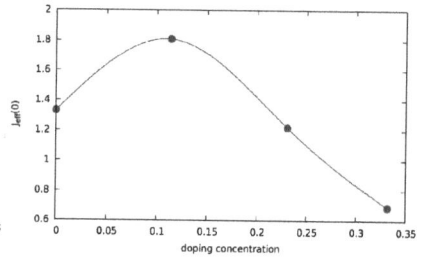

Figure 1d:

FIGURE 4.1   Exchange interaction strength vs. doping concentration in 1D; (a) antiferromagnetic (theoretical) (b) antiferromagnetic (experimental) (c) ferromagnetic (theoretical) (d) ferromagnetic (experimental).

Source: Reprinted with permission from the paper by Bhattacharjee, S., & Chaudhury, R., (2018). JLTP. With kind permission of Springer, 2018).

It is possible to define elementary excitations which are spatially local-ized to a narrow region in the direction normal to the surface, but are of extended type in directions parallel to the surface. Both quasi-particles and collective excitations can occur. The quasi-particles are basically electrons localized in surface electronic states. In similar ways, corresponding to the collective excitations of the lattice atoms, one can introduce "surface phonons" and "surface polaritons" as collective excitations of the surface layer. Moreover, there are collective excitations (plasma oscillations) of the electron gas near the surface, known as "surface plasmons," as opposed to "bulk plasmons" arising in the interior of the system.

We now explore the influence of the existence of a surface on the electronic energy spectrum. The surface we assume to be an idealized one. Furthermore, we take a one-dimensional model to start with. The surface then reduces to a point of discontinuity of the medium from perfectly periodic lattice to the vacuum. The lattice periodic potential can be represented as:

$$V(z) = V(z + ma) \qquad (4.8)$$

for $z < 0$, a being the lattice constant and;

$$V(z) = V_0 \qquad (4.9)$$

for $z > 0$ where, $V(z)$ is the potential experienced by an electron. The idea then is to solve the Schrodinger equation for this system and match the solutions at $z = 0$. Only the physically admissible solutions are considered. This leads to an exponentially decaying solution in space in the vacuum region and a band or Bloch function exhibiting a propagating character inside the solid. Detailed mathematical analysis shows that only with positive values of V, the matching of the solutions belonging to the interior of the solid and that from the outer region (vacuum) is possible. This generates an additional energy eigenvalue situated within the energy gap region between the bands. This corresponds to a quantum state in which the electron is spatially localized to a narrow region at the surface. This very discrete state is known as a "surface electronic state" for one-dimensional system (Madelung, 1996).

For three-dimensional system, the surface states form a new pattern. This is because of the fact, that for each value of the components of electron wave-vector parallel to the surface, we can expect a different position of the energy level of the surface electronic state. This leads to energy bands for the surface electronic states as well (Madelung, 1996).

The model for the lattice potential we have made use of above is unrealistic in several aspects. The first and foremost point is that a true surface is not a sharp transition from a perfect periodic potential to vacuum. Moreover, in most of the materials there are many other important issues like the presence of dangling bonds and adsorption layers on the surface.

There are several varieties of surface electronic states readily observed experimentally in real materials *viz.* mostly in semiconductors. They may arise due to the local distortions of the surface which originate from the presence of individual adsorbed atoms, incomplete adsorption layers, steps in the surface layer, etc. They are observed when the surface bands are absent. Some of the well-known states amongst them are "Shockley states" and "Tamn states" which play important roles in deciding the properties of the semiconductors (Shockley, 1939; Tamn, 1932). The Shockley states are the energy levels existing in the gaps or forbidden bands by the surface discontinuity of the Kronig-Penney model; whereas the Tamn states are calculated in the framework of a tight-binding Hamiltonian. There is also existence of "space charge layers" due to the exchange of electrons between the surface and the bulk electronic states (Bengtsson, 1999; Leung et al., 2003).

There are two other very important and interesting phenomena associated with surfaces. They are (i) "surface melting" and (ii) "surface reconstruction." Both of these involve the vibrations of the atoms on the surfaces. Surface melting is similar to bulk melting in that a condition like Lindemann Criterion is to be satisfied for the melting to take place. This happens comparatively easily on surfaces because of the presence of free or dangling bonds of the atoms on the surface. The idea of density wave freezing (Ramakrishnan and Yussouff, 1977, 1979) has also been used to understand the process of surface melting often (Tosatti et al., 1988, 1999).

The phenomenon of surface reconstruction on the other hand, takes place when a certain surface phonon branch gets highly softened. The softening itself arises due to anharmonicity or due to electron-phonon interaction. Once it happens, it is feasible for the surface atoms to rearrange themselves and form a new structure (Needs et al., 1991). This transition has also been modeled using the idea of topological phase transition, well known for 2D XY spin model in the field of magnetism (Kankaala et al., 1993). Experimental techniques such as X-ray or Neutron Diffraction and also atomic force microscopy (AFM) as well as scanning tunneling microscopy (STM) can detect occurrences of both of the above phenomena *viz.* surface melting and surface reconstruction (Sandy et al., 1991; Khaydukov et al., 2014; Oden et al., 1993). Besides, computer simulation techniques have also been used widely to study the above mentioned surface related phenomena (Wang et al., 2007).

Before we conclude, we throw some light on a class of novel systems called "topological insulator" which have been prepared by material scientists recently (Hasan et al., 2008). They have the strange property of behaving as an insulator inside the bulk and as a conductor like a metal at the surface. It has been established after thorough investigation that this puzzling behavior has essentially to do with the electronic structure in the presence of spin-orbit coupling, rather than many-body effects (Hasan and Kane, 2010). It is decided by the topological property of the electronic wave-function corresponding to the bands, which exhibits localized states inside the bulk and extended states on the surface, for example in systems like $Bi_{1-x}Se_x$. These materials are again classified as (i) Dirac insulator type and (ii) Weyl semi-metal type (Yan and Felser, 2016). Very interestingly, some of these systems after chemical doping turn into superconductors, as seen in $Bi_{1-x}Cu_xSe$. Investigations have also been going on both at the theoretical level and at the experimental one regarding the nature of the electronic tunneling current through a junction made up of a topological insulator and a conventional superconductor (Das, 2013; Zuo et al., 2016). These processes lead to the emergence of the role of so-called "Majorana Fermions," a rather novel concept in the realm of condensed matter physics (Qi and Zhang, 2011) (Figure 4.2).

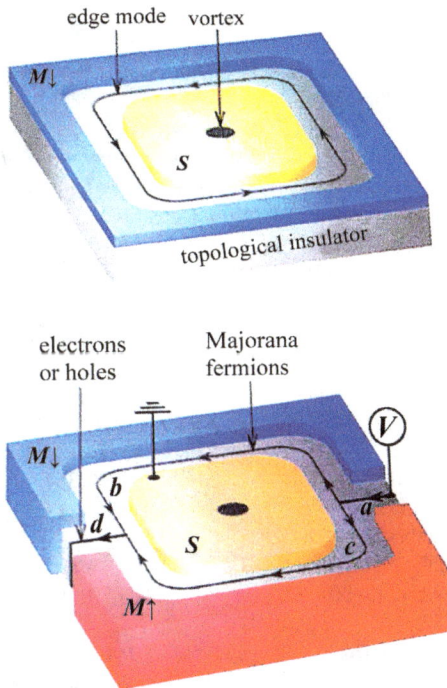

FIGURE 4.2   Emergence of "Majorana modes" in 3d topological insulators in proximity to ferromagnets with opposite polarizations and to a superconductor.

Source: Reproduced with permission from the paper by Qi, X. L., & Zhang, S. C., (2011). *Rev. Mod. Phys.* With kind permission of American Physical Society and authors. ©APS 2011.

KEYWORDS

- atomic force microscopy
- Landau's Fermi liquid
- Mott-Hubbard insulator
- random phase approximations
- Tomonaga-Luttinger liquid
- Wigner solid

REFERENCES

Atland, A., & Simons, B., (2006). *Quantum Field Theory in Condensed Matter.* Cambridge University Press, Cambridge.

Bhattacharjee, S., & Chaudhury, R., (2016). *Physica B., 500,* 133.

Bhattacharjee, S., & Chaudhury, R., (2018). *J. of Low Temp. Phys., 193,* 21.

Bhattacharjee, S., (2016). *J. Material Science and Engineering (Special Issue as Proceedings of Condensed Matter Physics Conference Held in Chicago in October 2016), 5,* 75.

Chaudhury, R., & Carbotte, J. P., (1993). *Solid State Communications, 85*(12), 1039.

Chaudhury, R., & Das, M. P., (2018). (Under preparation).

Chaudhury, R., & Gangopadhyay, D., (1995). *Mod. Phys. Lett. B., 9*(25), 1657.

Chaudhury, R., & Jha, S. S., (1984). *Pramana, 22*(5), 431.

Chaudhury, R., (1995). *Can. J. Phys., 73,* 497.

Chaudhury, R., (2009). *arXiv: 0901.1438v1 [cond-mat.supr-con].*

Das, M. P., (2013). Invited Talk at S. N. Bose Center, Kolkata, India.

Gell-Mann, M., & Brueckner, K., (1957). *Phys. Rev. B.,106,* 364.

Guiliani, G. F., & Vignale, G., (2005). *Quantum Theory of the Electron Liquid.* Cambridge University Press, Cambridge.

Hasan, M. Z., & Kane, C. L., (2010). *Rev. Mod. Phys., 82,* 3045.

Hsieh, D., Qian, D., Wray, L., Xia, Y., Hor, Y. S., Cava, R. J., & Hasan, M. Z., (2008). *Nature, 452*(7190), 970.

Jagla, E. A., Prestipino, S., &Tosatti, E., (1999). *Phys. Rev. Lett., 83,* 2753.

Kankaala, K., Ala-Nissila, T., & Ying, S. C., (1993). *Phys. Rev. B., 47*(4), 2333.

Khadukov, Y. N., Soltwedel, O., Marchenko, Y. A., Khaidukova, D. Y., Csik, A., Acarturk, T., Starke, U., Keller, T., Guglya, A. G., &Kazdayev, K. R., (2014). *Cond-Mat arXiv. org/1412.4356.*

Lindhard, K. J., (1954). *Dan. Viden. Sels. Medd., 28,* 8.

Luttinger, J. M., (1963). *J. Math. Phys., 4,* 1154.

Madelung, O., (1996). *Introduction to Solid State Physics* (pp. 425–430). Chapter 9, Springer-Verlag Berlin, Heidelberg.

Mahan, G. D., (1981). *Many Particle Physics.* Plenum Press.

Needs, R. J., Godfrey, M. J., & Mansfield, M., (1991). *Surface Science, 242,* 215.

Oden, P. I., Tao, N. J., & Lindsay, S. M., (1993). *Journal of Vacuum Science and Technology B., 11,* 137.

Pines, D., (1999). *Elementary Excitations in Solids.* Perseus Books Publishing, Massachusetts.

Qi, X. L., & Zhang, S. C., (2011). *Rev. Mod. Phys., 83,* 1057.

Ramakrishnan, T. V., & Yussouff, M., (1977). *Solid State Communications, 21,* 389; *Phys. Rev. B., 19,* 2775 (1979).

Roy, C. S., & Chaudhury, R., (2015). *Physica B., 465,* 60.

Roy, C. S., & Chaudhury, R., (2016). *Cond-mat arXiv. 1506.08373v3 [condmat.supr-con].* Cond-mat arXiv. 1807.11188v1 [cond-mat.supr-con] (2018).

Ruvalds, J., (1987). *Phys. Rev. B., 35,* 8869.

Sandy, A. R., Mochrie, S. G. J., Zehner, D. M., Huang, K. G., & Gibbs, D., (1991). *Phys. Rev. B., 43,* 4667.

Shastry, B. S., (1978). *Phys. Rev. B., 17*(1), 385.

Shastry, B. S., Jha, S. S., & Rajagopal, A. K., (1974). *Phys. Rev. B., 9*(4), 2000.

Shockley, W., (1939). *Phys. Rev., 56*(4), 317.

Stern, F., (1957). *Phys. Rev. Lett., 18,* 546.

Tamn, I., (1932). *Phys. Z. Soviet Union,1,* 733.

Tomonaga, S., (1950). *Prog. Theor. Phys., 5,* 544.

Trayanov, A., & Tosatti, E., (1988). *Phys. Rev. B., 38,* 6961.

Varma, C. M., (2015). *Cond-mat arXiv: 1601.03403v2 [cond-mat. str-el].*

Wang, Y., Hush, N. S., & Reimers, J. R., (2007). *Phys. Rev. B., 75,* 233416.

Yamada, K., (2004). *Electron Correlation in Metals.* Cambridge University Press, Cambridge.

Yan, B., & Felser, C., (2016). *arXiv: 1611.04182v2 [Cond-mat.mtrl-sci].*

Zuo, Z. W., Kang, D. N., Wang, Z. W., & Li, L., (2016). *arXiv: 1603.01361v2 [Cond-mat. Str-el].*

CHAPTER 5

Superconductivity

5.1 INTRODUCTION

The phenomenon of superconductivity is a manifestation of quantum coherence on a macroscopic scale. The phenomenon essentially is the vanishing of DC electrical resistance of materials at very low temperatures. It in fact describes a phase transition from a metallic state to a new phase where even the magnetic response is drastically different from that in the normal conducting state. Moreover, various thermodynamic properties also exhibit very different nature in these two phases. The phenomenon was discovered in 1911 by H. Kamerlingh Onnes (and Gilles Holst, one of his technical assistants) in the low-temperature physics laboratory at Leiden, while cooling solid *Hg* at liquid *He* temperature (Onnes, 1911; Delft, 2007). Until 1986 the superconductivity was experimentally observed only in various metals (including the organic ones) and metallic alloys. However, in the last two decades, many ceramic oxide systems have been synthesized which exhibit superconductivity at rather higher temperatures. This has opened up a new front in the research on superconductivity, both basic and applied.

The two very fundamental and universal properties of all superconductors are: (i) zero DC electrical resistance, and (ii) perfect diamagnetism. The second property is the well known Meissner-Ochsenfeld effect. The above two properties are in fact very much connected, as we will see later. Besides, the concept of electrical resistance itself will have to be looked at from a very different perspective than that used in a normal conductor. Apart from the above two fundamental properties, most of the superconductors possess some more characteristic features. They are the existence of a gap in the excitation spectrum and a definite functional dependence of the superconducting transition temperature on the mass of the ions of the lattice in the material. The second feature is called "isotope effect."

The superconductors may be classified into several categories in terms of some of the above properties. Based on the diamagnetic responses as a function of applied field strength, the superconductors are characterized as "*Type I*" and "*Type II*." The *Type I* superconductors exhibit complete orbital diamagnetism up to a field strength of H_c. Beyond this field, the system becomes a normal metal. The *Type II* superconductors, on the other hand, show complete Meissner screening up to field strength of H_c1. Thereafter, one observes an incomplete Meissner effect up to a maximum field limit of H_c2. Beyond this; again, the system turns into a normal metal. There is a third category of so-called "*Type 1.5*" proposed very recently (Moshchalkov et al., 2009; Brandt and Das, 2010). This is observed in certain two-band superconductors like MgB_2, where an apparent coexistence of both *Type-I* and *Type-II* behavior is found. Again, in terms of the existence or non-existence (vanishingly small magnitude) of the superconducting gap, the superconductors are called "gapped" or "gapless." Furthermore, on the basis of the symmetry properties of the gap function in momentum-space, the superconductors are further classified as "isotropic" or "anisotropic." Bringing in the role of microscopic mechanism, the superconductors obeying complete isotope effect are known as "phonon-based"; whereas the others are called "non-phonon dominated."

For the superconductors possessing gap, the electronic part of the specific heat $C_v^{el}(T)$ shows an exponential dependence on temperature in the superconducting phase experimentally. More precisely, $C_v^{el}(T)$ behaves approximately as $\exp\left(-\dfrac{\Delta(T)}{kT}\right)$ for $T < T_C$. Above T_C in the normal metallic phase, $C_v^{el}(T)$ is proportional to T, as usual. Thus, there is a discontinuity in specific heat at the superconducting transition temperature. This signals that this transition is a second-order phase transition. The function $\Delta(T)$ is itself a very non-trivial function of T and can be determined theoretically as well as experimentally. The more direct experimental proof of the existence of the superconducting gap comes from the irradiation of superconductors with electromagnetic waves. It is observed that there exists a threshold frequency of the electromagnetic radiation for the absorption to take place and that this frequency is given by $\dfrac{4\pi\Delta}{h}$.

Another very important and universal property of superconductors is "flux quantization." This is the quantization of the magnetic flux trapped inside a superconducting cylinder. It is experimentally observed that the flux is quantized in units of $\dfrac{hc}{2e}$. This effect can be understood on the basis of the property of perfect diamagnetism of superconductor and the electromagnetic gauge invariance. The occurrence of the quantity 2e in denominator is very significant, as it implies the charge carriers in the superconductors to be of a composite nature, consisting of a pair of electrons.

Based on these varieties of experimental results discussed above, several phenomenological theories were proposed at various times to extract some physical understanding of the phenomenon. The most well known amongst them are: (i) Gorter-Casimir (GC) two-fluid model, (ii) London-Pippard's (LP) theory, and (iii) Ginzburg-Landau (GL) theory (Schrieffer, 1983). All these theories hypothesize the existence of two fluid components in a superconductor. They are (i) a normal fluid component consisting of the normal electrons, the usual carriers in a metal, and (ii) a super-fluid component consisting of the superelectrons. The normal fluid is very similar to the normal metal; whereas the super-fluid consists of the condensate of electrons exhibiting radically different behavior from that of the normal fluid. It is assumed that at zero temperature the superconductor consists entirely of the super-fluid component; however, at T_c only the normal fluid component survives. At temperatures in between, the relative fraction of the above two components can be determined from these theories. The common starting point of both GL theory and GC theory is the difference in the free energy of the superconducting and normal phases, expressed either as a power series expansion in the relative fraction of the superfluid (in GL theory) or as a weighted sum of the contributions from the normal component and the super-fluid component (in GC theory). On the other hand, in the case of LP theory, the total charge current density in a superconductor is considered taking into account the contributions from both the components with equal weights. In a nutshell, the GC theory is based on the thermodynamic approach, the LP theory on the electro-dynamic approach and the GL theory is based on the general theory of second-order phase transition.

Before discussing very briefly about the above standard phenom-enological theories, let us look at the origin of the existence of a critical magnetic field in a superconductor, based on the general thermodynamic arguments. The perfect diamagnetism of a superconductor implies that the orbital magnetic energy of a superconductor in the presence of a magnetic field is enhanced by an amount $H^2/8\pi$ over the corresponding contribution in the normal state. This enhancement in Gibb's free energy must be compensated by the corresponding reduction in the Helmholtz free energy (due to the superconducting condensation) in order to have the superconducting phase comparatively more stable with respect to the normal phase. Therefore, if the strength of the applied magnetic field exceeds a certain critical value, the condensation energy cannot match the enhancement in the magnetic energy and the superconducting state becomes unstable compared to the normal phase.

In GC theory, the variational treatment on the proposed free energy model gives the equilibrium solution for the normal fluid fraction as a function of temperature. This is then used to calculate the condensation energy of the superconductor, which is also equal to $\dfrac{H_c^2}{8\pi}$. This equivalence leads to the following temperature dependence of the critical field (Schrieffer, 1983):

$$H_c(T) = H_0 \left[1 - \left(\frac{T^2}{T_c^2} \right) \right] \qquad (5.1)$$

This agrees very well with the experimental results. However, the predictions regarding $C_v^{el}(T)$ do not agree with results from most of the superconductors.

In LP theory, an attempt is made to reconcile the two fundamental properties of a superconductor, viz. infinite DC electrical conductivity and the Meissner effect. This requires the replacement of the standard constitutive equation for a normal metal viz. Ohm's Law with a corresponding new equation for the superconducting phase. This new equation relates the superfluid charge current density with vector potential generated from the applied magnetic field. When the equation is local, it is called Londons' equation. In the situation when this equation takes a more generalized non-local form, it is known as Pippard's equation. One of the important consequences of these equations is the emergence of the concepts of (i) 'penetration depth' or 'screening length' and (ii) 'coherence length.' It turns out from detailed analysis even using London's equation, that the magnetic induction field decays exponentially inside a superconductor and that the complete orbital diamagnetism appears only in the bulk of the superconductor. This characteristic decay length is known as "London's Penetration Depth" or "Meissner Screening Length." Similarly, in Pippard's equation, the vector potential is averaged over a length scale postulated to be the average size of the super-fluid particles. This length is called "Superconducting Coherence Length." These two lengths discussed above, play very important roles in deciding various properties and responses of superconductors both theoretically and experimentally. Moreover, in combination with the results from GC theory, one can predict the temperature dependence of the penetration depth as:

$$\lambda_L(T) = \frac{\lambda_L(0)}{\left[1 - \left(\frac{T}{T_c}\right)^4\right]^{0.5}}$$

(5.2)

where, the zero-temperature penetration depth $\lambda_L(0)$ is given by $\frac{c}{\omega_p}$; ω_p being the plasma frequency in the normal phase of the superconductor.

Furthermore, the magnitude of the ratio of the penetration depth and the coherence length, known as the "Abrikosov-Gorkov parameter," decides the nature of the superconductor as type-I or type-II (Schrieffer, 1983; Abrikosov, Gorkov, and Dzyaloshinskii, 1975).

The GL theory is a very general one but a powerful one dealing with the entire class of second-order phase transitions in statistical mechanics (Mahan, 1981). Mostly the situation in the ordered phase at a temperature close to transition temperature T_c is analyzed. Thus, the power series expansion is generally terminated at the 4th order in the order parameter. Moreover, the cases both in the presence and in the absence of the external magnetic field are considered. Once again, the variational scheme is applied and the expression for the equilibrium solution for the order parameter viz. the relative fraction of the superfluid component is obtained. This can also be combined consistently with the corresponding results from the earlier mentioned phenomenological theories of GC and LP. These resulting equations can then be used to determine the expansion coefficients (also known as GL coefficients) in GL theory. The situation with the presence of an external magnetic field can also be tackled in a similar fashion. The analysis with the magnetic field can be suitably extended to study the spatial distribution of the magnetic induction field and the superconducting order parameter inside a type II superconductor. In particular, this scheme is widely used to predict the nature and the properties of the so-called "vortex state," a particular configuration involving the coexistence of the so-called the "normal" and the "superconducting" regions, in the magnetic field regime lying between the lower and the upper critical fields. This procedure also brings out very explicitly the two characteristic length scales *viz.* penetration depth and coherence length in the calculational results for the superconducting order parameter and the magnetic field distribution.

So far, we have only described in a very sketchy way the phenomenological theories related to the conventional superconductors. Now we turn to the most challenging aspects, *viz.* various proposed microscopic theories for superconductivity. We first discuss about the conventional systems *viz.* metals and metallic alloys. Some of the key questions which are expected to be addressed by the microscopic theories are: (i) the mechanism for the formation of the super-electrons *viz.* a composite of two electrons, (ii) the origin of the existence of the superconducting gap, and (iii) the origin of isotope effect and the last but not least, the underlying process causing the metal-superconductor transition.

5.2   MICROSCOPIC THEORIES

One of the earliest attempts came from Frohlich, in particular for the explanation of the "isotope effect"(Frohlich, 1950; Schrieffer, 1983). He guessed that lattice is involved in some way in bringing about superconductivity. His further detailed analysis involving a full Hamiltonian consisting of the electron-lattice system brought out very clearly the possibility of an electron-electron effective attractive interaction arising from the exchange of phonons. This issue has been discussed at some length earlier in Chapter 3. The next important milestone was Cooper's work on the formation of the quasi-bound electron pair in the background of a Fermi Sea. The detailed quantum mechanical calculation shows that two electrons experiencing an attractive interaction of arbitrarily small magnitude can form such a quasi-bound pair with respect to the Fermi surface. The binding energy of such a pair was also calculated as a function of the magnitude of the attractive interaction. This very important result leads to two major candidates for the microscopic theories *viz.* (i) Bardeen-Cooper-Schrieffer (BCS) theory and (ii) Scafroth-Blatt-Butler (SBB) theory, motivated by the works of N.N. Bogoliubov and collaborators. The BCS theory makes use of Frohlich's idea and Cooper's results and starts from a many-body Hamiltonian involving fermions in states exhibiting "pairing correlations." The SBB theory, on the other hand, considers the "paired fermions" to be bosons and deals with an effective bosonic Hamiltonian.

We now first present a broad description of "Cooper's one pair problem." Then later in some details, we highlight the principal features of the BCS theory as well as discuss its various consequences in quantitative terms, including the behavior of the superconducting gap and the expression for the superconducting transition temperature. We furthermore stress on the success of the BCS theory by making comparisons between the theoretical predictions and the experimental results from various conventional and non-conventional systems. Regarding the SBB theory, we touch upon its major merits and demerits keeping in perspective various types of superconductors including the exotic ones.

5.3   COOPER'S ONE PAIR PROBLEM

Cooper considered a pair of electrons in the singlet spin state and with the orbital configuration of the type $k + \frac{q}{2}, -k + \frac{q}{2}$ in the vicinity of the Fermi surface, where the vectors k and q are the momenta or more precisely the wave vectors. Thus, the pair $(\uparrow, \downarrow)$ has a center of mass momentum $hq/2\pi$. Later Cooper considers the pair with $q = 0$ only, where the two electrons occupy the time-reversed states. Then the anti-symmetrized orbital wave function of the pair is incorporated in Schrodinger's time-independent equation with the interaction potential being only space dependent. This is expressed as:

$$[H_0 + V(r_1, r_2)]\psi_{orb}(r_1, r_2) = W\psi_{orb}(r_1, r_2) \qquad (5.3)$$

where, H_0 is the kinetic energy operator of the electron pair, which is equal to the sum of the kinetic energies of the two electrons. In the general case, the orbital wave function of the pair is expressed as:

$$\psi_{orb}(r_1, r_2) = \phi_q(\rho)\exp(iq \cdot R) \qquad (5.4)$$

where, the function φ defined in the relative coordinate space, can be expanded as:

$$\phi_q(\rho) = \sum_k \left[a_{k+q} \exp(ik \cdot \rho) + a_{-k+q} \exp(-ik \cdot \rho) \right] \qquad (5.5)$$

For the $q = 0$ pairs, only the first term is sufficient and is retained in the above expression. Now performing a scalar product with the plane wave function $exp(ik \cdot \rho)$ on the two sides of Schrodinger's Eqn. (5.3), we arrive at the following equation:

$$\left(W - 2\epsilon_k \right) a_k = \sum_{k'} V_{k,k'} a_{k'} \qquad (5.6)$$

where, $V_{k,k'}$ is the Fourier transform of the attractive interaction in real space $V(r_1, r_2)$.

Following Cooper, we model the real space interaction to be a contact interaction and attractive when the electron energies are within a shell of width of the order of Debye energy above the Fermi level and zero otherwise. Then, Eqn. (5.6) leads to the following equation corresponding to the quasi-bound state for the electron-pair:

$$\frac{1}{U} = \sum_k \frac{1}{|W| + 2\epsilon_k} \qquad (5.7)$$

where, U is the magnitude of the attractive interaction, W is the energy eigenvalue for the bound state and hence is negative when the zero of the energy is shifted to $2E_F$, with E_F being the Fermi energy. The summation over k is restricted to satisfy the requirement that the electron energy, as measured from the Fermi energy, is less than the cutoff energy (generally of the order of Debye energy for phonon mediated attraction). Choosing the zero of the energy at E_F and measuring the single-particle and the two-particle energies appropriately, we get the following equation for the binding energy of the pair:

$$E_B = \frac{2\hbar\omega_c}{2\pi \left(\exp\left[\dfrac{2}{g} \right] - 1 \right)} \qquad (5.8)$$

where, the binding energy E_B is equal to $|W|$, ω_c is the cutoff frequency corresponding to the boson mediating the attractive pairing interaction and the attractive coupling constant g is given as $N(0)U$, where $N(0)$ is the single-particle density of states at the Fermi energy for one kind of spin. This quasi-bound pair is known as "Cooper Pair."

The above analysis shows that a very large number of such bound pairs may be formed near the Fermi surface. This implies that the Fermi Sea becomes unstable when an arbitrarily small attractive interaction is introduced between the electrons. This is essentially the origin of super-conducting pairing in an ideal metal described by a simple Fermi Sea in the continuum limit, neglecting the effects of the discrete lattice.

The above Eqn. (5.8) can now be examined under various conditions. In the weak-coupling limit $viz.$ $g << 1$, the binding energy takes the following approximate form:

$$E_B^{wc} = \frac{h\omega_c}{\pi} \exp\left[-\frac{2}{g} \right]$$ (5.9)

In the strong coupling limit, i.e., $g >> 1$ on the other hand, we have:

$$E_B^{sc} = \frac{gh\omega_c}{2\pi}$$ (5.10)

For finite q-pairing, one can show by detailed calculation that the constraints on the magnitude of the electron energies lying within the so-called Debye cutoff above the Fermi level, for the formation of quasi-bound state lead to the reduction of the binding energy. It is seen that the relation between the binding energy corresponding to the finite center of mass momentum E_B^q and that corresponding to the zero centers of mass momentum E_B^0 are related by (Schrieffer, 1983):

$$E_B^q = E_B^0 - v_F q \frac{h}{2\pi} \quad \text{(for weak coupling)}$$ (5.11)

and

$$E_B^q = E_B^0 - g v_F q \frac{h}{2\pi} \quad \text{(for strong coupling)}$$ (5.12)

where v_F being the Fermi velocity of the electrons.

The above Eqn. (5.11) for binding energy reduction in the case of weak coupling shows that the binding energy goes to zero for a maximum value of q denoted by q_{max}. Making use of the phenomenological consideration viz. E_B^0 is of the order of kT_c, leads us to the interesting relation that the q_{max} is of the order of $K_F \dfrac{T_c}{T_F}$. The reciprocal of this q_{max} is an estimate for "superconducting coherence length," as referred to earlier in the context of phenomenological theory and is denoted generally by ξ. Physically this is an outcome of the uncertainty principle, as this length describes the size of a Cooper pair and the quantity $\dfrac{hq_{max}}{2\pi}$ is the spread in the center of mass momentum of the pair. For weak coupling superconductors, the coherence length is generally of the order of 10^4 lattice spacings and finds experimental support from the measurements in the conventional super-conductors (Schrieffer, 1983).

This important work of Cooper laid the groundwork for constructing a true microscopic theory for superconductivity. Furthermore, Frohlich's treatment had already provided a source for the attractive interaction. As mentioned before, two different types of microscopic theories emerged. The proponents of the first one are Scaffroth, Blatt and Butler (Blatt, 1964) and the theory was put forward in 1957. It treats Cooper pairs as Bosons and considers the superconducting transition as a kind of Bose-Einstein condensation (BEC) phenomenon. More precisely, it considers T_c as the temperature at which almost all (in principle a finite fraction) the Cooper pairs occupy the $q = 0$ state. The second theory was proposed by Bardeen, Cooper, and Schrieffer a little later, in the same year, i.e., 1957 (Schrieffer, 1983). This theory keeps the Fermionic nature of the electrons forming the Cooper pairs intact and constructs a Fermionic many-body Hamiltonian. The many-body term however has electrons interacting only in the Cooper pair configurations. In this formalism, T_c represents the temperature below which the thermodynamic average of the number of Cooper pairs becomes non-zero. It also brings out the microscopic definition of superconducting gap function.

We now present the major features of both of these theories in some details. For the BCS theory, we carry out a semi-qualitative-semi-quantitative overview. Regarding the SBB theory, we only present a qualitative picture. We also make some critical comments about the merits and demerits of these two theories in the perspective of the experimental results from conventional and non-conventional superconductors.

5.4 BCS THEORY

The BCS theory proposes a new Hamiltonian for describing the super-conducting phase. At $T = 0$, i.e., in the superconducting ground state, all the electrons are in the paired configurations and the system is described by the reduced Hamiltonian involving only the Cooper Pair operators. At finite temperature, however, the more generalized Hamiltonian, involving both single electron operators as well as pair operators, are to be used. It is also equally interesting to look at the nature and the structure of the BCS ground state itself. The BCS ansatz for the ground state is that the ground state is a direct product of all the paired states, each paired state being completely occupied or completely empty. The amplitudes corresponding to these occupancies are determined variationally.

Mathematically, this is expressed as:

$$|\psi\rangle_G = \Pi_k \left[u_k + v_k b_k^+ \right] |0\rangle \qquad (5.13)$$

where, the ket vectors on the left and right-hand sides represent the BCS ground state and the filled Fermi Sea, respectively. Inside the Fermi Sea, the electrons are passive and do not take part in Cooper pairing. From now onwards for simplicity, we will drop the bold sign for vectors. The complex quantities u_k and v_k represent the probability amplitudes corresponding to the zero occupancies and the full occupancy of the Cooper pair state respectively. The operator b_k^+ is the Cooper pair creation operator and is given by:

$$b_k^+ = c_{k\uparrow}^+ c_{-k\downarrow}^+ \qquad (5.14)$$

A similar equation is satisfied by the Cooper pair destruction operator b_k. The probability amplitudes obey the constraint;

$$|u_k|^2 + |v_k|^2 = 1 \qquad (5.15)$$

Physically the above-chosen structure for the BCS ground state is based on the idea that the Cooper pair states give the lowest energy states for the two-electron configuration with the inter-electron interaction being attractive and this interaction having non-vanishing matrix elements only between two Cooper pair states. One of the very interesting consequences of the above structure is that the total number of particles is not fixed in the ground state of the system, can be seen easily by the operation of $\sum_k b_k^+ b_k |\psi\rangle_G$.

Let us now focus our attention to the BCS Hamiltonian. At zero temperature, the Hamiltonian takes the simple form:

$$H^0 = \sum_k 2\,\epsilon_k\, b_k^+ b_k + \sum_{k,k'} V_{k',k} b_k^+ b_k \qquad (5.16)$$

where, $V_{k'\uparrow,-k'\downarrow}$ is the inter-electron attractive interaction matrix element (and therefore is of negative magnitude) describing a transition between the Cooper pair states $(k\uparrow,-k\downarrow)$ to $(k'\uparrow,-k'\downarrow)$. The first term on the right-hand side is the kinetic energy associated with the pairs and the second term, the interaction term, may in fact be looked upon as the transfer term for the pairs from one state to another. The variational approach with the objective of minimizing the ground state energy and maintaining the constraint that the expectation value of the particle number operator in the ground state is equal to the actual number of particles in the system, we can determine the constants u_k and v_k. This leads us to the following results for the above quantities:

$$|u_k|^2 = \frac{1}{2}\left[1 + \frac{\xi_k}{E_k}\right];\ |v_k|^2 = \frac{1}{2}\left[1 - \frac{\xi_k}{E_k}\right] \qquad (5.17)$$

where;

$$\xi_k = \epsilon_k - \mu \qquad (5.18)$$

μ being the chemical potential and;

$$E_k = \left[\xi_k^2 + |\Delta_k|^2\right]^{\frac{1}{2}} \qquad (5.19)$$

with the zero temperature BCS gap function Δ_k being defined as:

$$\Delta_k = -\sum_{k'} V_{k',k} \langle b_{k'}^+ \rangle \qquad (5.20)$$

where, the average implies expectation value in the ground state. It may be noted that in general Δ_k is a complex quantity. The quantity E_k has a special physical significance, as we will see later. Furthermore, it follows that the gap function satisfies a self-consistent equation given below:

$$\Delta_k = -\sum_{k'} V_{k,k'} \frac{\Delta_{k'}}{2E_{k'}} \qquad (5.21)$$

The incorporation of these variationally obtained parameters in the formal expression for the expectation value of the Hamiltonian in the BCS ground state yields the BCS ground state energy. We do not attempt to present the expression here, as it is extremely long (Schrieffer, 1983).

The special property of the BCS ground state of possessing variable number of particles is in fact quite important from the standpoint of statistical mechanics. It is a manifestation of "symmetry breaking" as the variable number of particles implies the fixation of the phase of the system; whereas the Hamiltonian given by Eqn. (5.16) is invariant under introduction of any arbitrary phase and commutes with the total particle number operator.

Let us now turn our attention to the more generalized case *viz.* that of finite temperature. Here the BCS Hamiltonian takes the form (Schrieffer, 1983):

$$H_{gen} = \sum_{k,\sigma} \epsilon_k \, c_{k,\sigma}^+ c_{k,\sigma} + \sum_{k,k'} V_{k,k'} b_k^+ b_k \qquad (5.22)$$

As explained before, the above Hamiltonian now has both single electron contributions as well as Cooper pair contributions. This is physically understandable, as a superconductor at finite temperature can be represented as a mixture of "normal fluid" and "super-fluid" components, according to the scenario of the Two-Fluid model of GC, unlike the zero temperature case where only the super-fluid part survives.

Just like as in the zero temperature case, one can again define a finite temperature gap function by the Eqn. (5.20), where the expectation value of the pair creation operator will now have to be understood as a "thermodynamic average." This gap function is the "order parameter" for the superconducting phase. It can be argued that in general the BCS gap is a complex quantity. However, often the phase of this gap function is chosen to be zero, so that it becomes a real quantity. The physical implication of this will be discussed later.

We now use the standard techniques to handle the above many-body Hamiltonian given by Eqn. (5.22). The simplest thing to try is the Hartree-Fock or Mean Field technique. One decouples the interaction term in such a way so as to extract the gap parameter and at the same time maintaining the Hermitian character of the Hamiltonian intact. This however turns the mean-field Hamiltonian non-commuting with the total particle number operator, unlike the original Hamiltonian. Thus, it becomes necessary to carry out the statistical mechanical analysis and all the relevant calculations for thermo-dynamic properties within the framework of the Grand Canonical Ensemble. The chemical potential (μ) automatically occurring in this formalism, can be determined from the constraint that the expectation value of the total particle number operator evaluated using the density matrix, is exactly equal to the total number of electrons observed in the system physically.

It should be highlighted that the kind of mean-field treatment described above, leads to "anomalous averages," i.e., average over a pair of particle creation operators or over a pair of hole creation operators, as propor-tional to the order parameters. This is quite distinct from the conventional averages over a combination of a particle creation and a hole creation operators, which plays the roles of the order parameters in the problems of magnetism and charge density wave (CDW) formations.

The mean-field BCS Hamiltonian is bi-linear in fermion operators. Hence, it can be diagonalized by Bogoliubov-Velatin (BV) transforma-tion. This leads to the emergence of quasi-particle operators which are fermionic in nature; however, they are a suitable linear combination of the original fermion operators. They in fact describe the stable single-particle excitations from the breaking up of Cooper pairs. The energy E_k associated with these quasi-particle operators will appear as eigenvalues of the mean-field Hamiltonian. The energy versus wave-vector dispersion of the quasi-particles however differs drastically from those of the usual fermionic carriers of the normal metal, in that they exhibit an energy gap. This is clearly seen in Eqn. (5.19). This energy gap is the same as the wave-vector dependent BCS gap parameter.

Mathematically, all these are very similar in structure to the zero temperature case described earlier. The self-consistent gap equation at the finite temperature is almost identical to its zero temperature counterparts

(Eqn. (5.21)) except that it carries an additional factor of $\tanh\dfrac{\beta E_k}{2}$ in the right-hand side of the equation.

The structure of these quasi-particle operators is of the following type:

$$\gamma_{k\uparrow}^{+} = u_k c_{k\uparrow}^{+} - v_k c_{-k\downarrow} \qquad (5.23)$$

$$\gamma_{-k\downarrow} = v_k c_{k\uparrow}^{+} + u_k c_{-k\downarrow} \qquad (5.24)$$

The diagonalized Hamiltonian then takes the form:

$$H_{gen} = \text{constant}.I + \sum_k E_k \left[\gamma_{k\uparrow}^{+}\gamma_{k\uparrow} + \gamma_{-k\downarrow}^{+}\gamma_{-k\downarrow} \right] \qquad (5.25)$$

where, I stands for identity operator and the constant-coefficient signifies the ground state energy corresponding to this Hamiltonian. The ground state is the BCS ground state referred to earlier and the ground state energy is the variationally obtained energy.

The solution to the finite temperature gap equation describes the temperature dependence of the superconducting gap. In order to solve this equation, one has to model the interaction function $V_{k,k'}$. The earliest modeling of BCS was based on a 1-square model. In this model, the interaction function was taken to be a function of $|k|$ and $|k'|$ only and is finite only within a range of the single-particle energies (when measured from the Fermi energy) below a cutoff but vanishes outside. Moreover, the magnitude of this cutoff energy is taken to be of the order of Debye energy.

Mathematically this 1-square well model is described as:

$$V_{k,k'} = -V \qquad (5.26)$$

with $V > 0$ when both $|\xi_k| < \omega_c$ and $|\xi_{k'}| < \omega_c$. Otherwise, the interaction matrix element vanishes. The parameter ω_c is the cutoff.

Incorporating this form of the interaction potential in the gap equation, one readily sees that the gap function also takes the form of a square well. More precisely, the gap function Δ_k becomes a non-vanishing constant Δ_0 only when $|\xi_k| < \omega_c$. Otherwise, it becomes zero. This leads to the non-linear temperature-dependent self-consistent gap equation. This equation is satisfied by the function $\Delta_0(\beta)$. At zero temperature, it follows that the gap parameter obeys the same equation as the Cooper pair binding energy, in the weak coupling limit (see Eqn. (5.9)).

In order to determine T_c, one has to linearize the finite temperature gap equation. The equation then takes the form:

$$\frac{1}{g} = \int_0^{\omega c} \frac{d\xi}{\xi} \tan h \left[\frac{\xi}{2kT_c} \right] \tag{5.27}$$

For the conventional low-temperature superconductors, the parameter $\beta_c \omega_c >> 1$. In this limit, the equation for T_c takes the form:

$$T_c = 1.14\theta_c \exp\left[\frac{-1}{g} \right] \tag{5.28}$$

where, θ_c is the temperature equivalent of the cutoff frequency and is of the order of Debye temperature for phonon mediated attractive pairing interaction.

The above equation viz. Eqn. (5.28) is the celebrated BCS equation for T_c in the weak coupling limit. The equation shows the qualitative dependence of T_c on the cutoff frequency as well as on the magnitude of the attractive coupling constant. The cutoff frequency itself is set according to the microscopic mechanism for superconductivity viz. whether it is phononic or electronic. For the conventional superconducting materials, the phonon being the mediator of attraction, the cutoff frequency of the attractive well can be taken to be of the order of Debye frequency. Therefore, the superconducting transition temperature can in principle be raised if we invoke the exchange of bosons with higher characteristic energy than those of the phonons. For the superconductors driven by the phonon-based mechanism, the above equation for T_c exhibits the "isotope effect," as the cutoff frequency (proportional to the Debye frequency) is expected to vary as the inverse square root of lattice ionic mass.

The dependences of T_c on the cutoff frequency and the coupling constant are very similar to the ones seen for the zero temperature gaps and the Cooper pair binding energy in the weak coupling limit, as seen from the Eqns. (5.9) and (5.28). The transition temperature grows with the coupling constant in the non-analytic fashion (see Eqn. (5.27)) in the weak coupling regime *viz.* $g \ll 1$. The increase of T_c with both the cutoff frequency and the attractive coupling constant can be a bit tricky, however, with a detailed phonon spectrum in real materials. This is because of the fact that often in the case of phonon exchange based mechanism, the magnitude of the coupling constant goes down when the average phonon frequency goes up. This causes the T_c to exhibit a maximum as a function of g (Ginzburg and Kirzhnits, 1982).

The Eqn. (5.28) for T_c, derived under the 1-square well model approximation, is a bit simplistic, as it neglects the ever-important repulsive Coulomb interaction operating between the electrons. The more complete modeling thus should include both phonon-induced attractive interaction as well as the universal repulsive Coulomb interaction between the electrons. This is achieved by the 2-square well model. Besides the attractive well with the cutoff frequency of magnitude ω_c, one introduces a repulsive well with cutoff frequency of magnitude $\omega_F = \dfrac{2 \pi E_F}{h}$. It can be shown by detailed algebra that in this case, the Eqn. (5.27) for T_c gets modified by the appearance of $\tilde{g}$ in the place of g in the right-hand side where (Ginzburg and Kirzhnits, 1982):

$$\tilde{g} = N(0) \left[|V_{ph}| - \frac{|V_{Coul.}|}{S_c} \right] \tag{5.29}$$

with V_{ph} being the phonon induced attractive interaction, $V_{Coul.}$ being the repulsive Coulombic interaction and S_c is the Coulomb suppression factor given by:

$$S_c = 1 + N(0) |V_{Coul.}| \log \left[\frac{\omega_F}{\omega_c} \right] \tag{5.30}$$

The 2-square well model is a much more realistic model for superconductor, compared to the originally proposed 1-square well one. Another important property of 2-square well model is that it weakens the isotope exponent even when the microscopic mechanism for superconductivity arises entirely from the phonons. This is because the isotope exponent defined as $\dfrac{-d \log T_c}{d \log M}$ deviates from the ideal value of 0.5 (seen in the case of pure 1-square well model with frequency spectrum independent of g) in the presence of the repulsive Coulomb interaction, as can be verified from the modified equation for T_c (involving the new coupling constant $\tilde{g}$). This has a lot of significance for the experiments, as very often the observation of isotope exponent less than 0.5 is taken as an evidence for a non-phonon mechanism for superconductivity operating in the material.

The analysis so far has been restricted to the situation with weak electron-phonon coupling. In the case of strong electron-phonon coupling, the electron wave functions get highly renormalized. In this case, the explicit frequency dependence of the effective electron-electron pairing interaction, known as the retardation effect and the damping of the real or thermal phonons becomes very important. These important many-body effects have been incorporated in the scheme developed by G. M. Eliashberg and formalism for the calculation of T_c has been developed (Schrieffer, 1999; Eliashberg, 1960). This scheme has later been simplified by McMillan to obtain an approximate equation for T_c in the intermediate coupling regime (McMillan, 1968).

The Eliashberg equations deal with two kinds of quantities *viz.* single electron (fermion) self-energy and pairing (Cooper pair) self-energy. These can be expressed in terms of the frequency-dependent renormalization function $Z(i\omega_n)$ and the superconducting gap function $\Delta(i\omega_n)$ in the imaginary frequency, i.e., Matsubara representation (Abrikosov, Gorkov, and Dzyaloshinskii, 1961). The frequency-dependent effective electron-electron attractive interaction mediated through phonon or some other boson exchange mechanism ($\lambda(i\omega_m - i\omega_n)$) and the direct Coulomb repulsion ($\mu^*(\omega_c)$) also occur in these equations.

The equations are of the following form:

$$\Delta(i\omega_n)Z(i\omega_n) = \pi T \sum_m \left[\lambda(i\omega_m - i\omega_n) - \mu^*(\omega_c)\theta(\omega_c - |\omega_m|) \right] \frac{\Delta(i\omega_m) - \Delta(i\omega_m)}{\left[\omega_m^2 + \Delta^2(i\omega_m) \right]^{\frac{1}{2}}}$$

(5.31)

and

$$Z(i\omega_n) = 1 + \frac{\pi T}{\omega_n} \sum_m \lambda(i\omega_m - i\omega_n) \frac{\omega_n}{\left[\omega_m^2 + \Delta^2(i\omega_m) \right]^{\frac{1}{2}}}$$

(5.32)

where, T is the temperature and the attractive interaction kernel $\lambda(i\omega_m - i\omega_n)$ is related to the electron-boson spectral function, known also as the Eliashberg function, $\alpha^2 F(\omega)$ in the following way:

$$\lambda\left(i\omega_m - i\omega_n\right) = 2\int_0^\infty \frac{d\omega\omega\alpha^2 F\left(\omega\right)}{\omega^2 + \left(\omega_n - \omega_m\right)^2}$$ (5.33)

with $i\omega_n = i\pi T \, (2n - 1)$ is the nth Matsubara frequency corresponding to the fermions.

Simplification of the above equations yields the well known McMillan's equation (Carbotte, 1990):

$$T_c = \frac{\omega_{av}}{1.2}\exp\left[-\frac{1.04\left(1+\lambda\right)}{\lambda - \mu^*\left(1+0.62\lambda\right)}\right]$$ (5.34)

where, ω_{av} denotes the average boson frequency and λ represents the attractive coupling constant (denoted by g earlier in the context of the BCS theory) or the mass renormalization parameter. In the weak coupling regime *viz.* $\lambda \ll 1$, the above equation reduces to the BCS equation for T_c.

5.5 CONSEQUENCES AND APPLICATIONS OF BCS PAIRING THEORY

The concept of electron-electron pairing in the BCS theory leads to the important physical quantity called the "coherence factor." These factors originate from the BV transformation coefficients relating to the quasi-particle operators in the superconducting state to the normal electron operators. These factors appear in the expressions for various microscopic properties in the superconducting phase. Some of the well known and useful properties are (i) acoustic attenuation rate, (ii) NMR relaxation rate, and (iii) microwave (MW) absorption spectra. The coherence factors corresponding to these 3 properties are distinct from each other. The reason for that lies in the underlying microscopic processes. The normal electrons physically couple to the external probes like acoustic phonons, the nuclear spins, or electromagnetic waves through various types of interactions, spin-dependent, or spin-independent. The normal electron states however are not stable states quantum mechanically in a superconductor, unlike a metal. Therefore, we need to look upon the normal electron creation and destruction operators as a linear superposition of the Bogoliubov quasi-particle operators. Thus the perturbation Hamiltonian, depending on the form of the interaction, will now consist of various terms describing the coupling of the quasi-particle operators to the operators representing the external probes, with coefficients related to BV coefficients. The rate calculations for the above processes by means of Fermi Golden Rule will, therefore, contain these coefficients and the final expressions will exhibit various functional combinations of them implying interference effect between various processes. These various functional combinations of BV coefficients are the so-called "coherence factors." They play very vital roles in deciding the nature of the experimental results in going from the normal to the superconducting phase. In fact, the excellent agreement between the theoretical predictions and the experimental results for the conventional superconductors indicates the success of the BCS pairing theory.

There are several more direct evidences and applications of the BCS pairing theory. They are (i) Andreev reflection experiment, (ii) Josephson tunneling and single-particle tunneling, and (iii) superconducting quantum interference device (SQUID). In the Andreev reflection experiment, a super-conductor is bombarded with an electron beam and it is found that holes are emitted from the superconductor. This is a direct evidence of the pairing theory, as the injected electrons tend to get paired up with the electrons inside the material and as a consequence, the holes are emitted out.

The superconducting tunneling is of two types: (i) single-particle (or quasi-particle) tunneling, and (ii) pair tunneling or Josephson tunneling. The first process involves the superconductor-normal metal (SN) junction. The second one relates to the system consisting of two superconductors separated by an insulating barrier (SIS). In the former situation, an electron from a normal metal will have to occupy a quasi-particle state in the superconductor and hence will have to overcome an energy gap. This is reflected in the current-voltage characteristics by the existence of a threshold voltage. This process is known as single-particle tunneling effect. Using the Fermi Golden Rule, the entire current-voltage characteristics can be calculated theoretically for the SN junction. This tunneling current can be shown to be proportional to the Bogoliubov quasi-particle density of states (Carbotte, 1990).

In the case of SIS, Cooper pairs can tunnel coherently from the left superconductor to the right superconductor through the insulating barrier, provided the barrier is thin. This leads to an electrical current even in the absence of any applied external voltage. This effect is known as the DC Josephson effect. In the presence of an applied DC voltage, however, the Cooper pair wave function undergoes a time evolution according to the Schrodinger equation. This leads to a temporal oscillation of the pair current. This effect is called the AC Josephson effect. The frequency of this oscillation is proportional to the magnitude of the applied voltage. The experiments involving both SN junction and SIS configuration show very good agreement with the theoretical results derived on the basis of the BCS pairing theory (Carbotte, 1990).

The device SQUID works on the principle of Josephson junction. It consists of a superconducting ring with two Josephson junctions separated by a static magnetic field. The presence of the finite magnetic flux causes a phase difference to develop between the wave functions corresponding to the Cooper pairs tunneling through the two junctions. This leads to an interference between the wave functions corresponding to these different paths and produce a fringe pattern of the net tunneling current. The interference pattern observed in the tunneling current follows the predictions of the pairing theory.

The concepts of Josephson junction and SQUID find a lot of practical applications. The electrical transport in polycrystalline superconductors takes place through the natural formation of weak-link Josephson junctions inside the system. The SQUID is often used to measure magnetic fields of very small magnitude, where the conventional experimental technique fails. Apart from these, in medical science SQUID is used to map out the spatial distribution pattern of the internal magnetic field in human brains in the various brain scan techniques to detect neurological diseases.

Very recently, it has been shown that for the BCS superconductors in 3d, the electronic polarizability function exhibits a singularity very similar to the well-known Kohn singularity found in normal LFL metals, in the limit of superconducting coherence length becoming very large or very small compared to the lattice parameter (Chaudhury and Das, 2010, 2013). Furthermore, this leads to a singularity in phonon dispersion function as well at phonon energy equal to twice the superconducting gap, analogous to Kohn Anomaly observed in normal metals. This finds very good experimental support from elemental 3d superconductors like Pb and Nb, both of which also possess a phonon driven pairing mechanism (Aynajian et al., 2008). This motivated further theoretical analysis bringing out a self-consistent solution for an equation involving phonon spectra and electron-phonon coupling constant at the Kohn Anomaly point, based on the BCS equations for T_c and zero temperature gap (Chaudhury and Das, 2013). The anomalous behavior of the sound wave velocity indicating Kohn Anomaly from the experimental phonon dispersion curve in superconducting Pb, is beautifully presented in Aynajian et al. (2008) (Figure 2(B), p. 1510).

5.6   SBB THEORY

As briefly touched upon earlier, SBB theory treats the electron pairs formed by the effective attractive interaction, as bosons. Furthermore, the superconducting transition is looked upon as BEC of these composite bosons, similar to the superfluidity of He^4. There were however serious objections against this proposal, both from the theoretical viewpoint as well as experimental observations. The theoretical criticisms stem from the fact that the fermionic pair operators (creation and destruction) do not obey the Bosonic algebra exactly (Schrieffer, 1983). The more striking contradiction is from the experimental observations of the specific heat near T_c. The behavior of the electronic part of C_v in conventional superconductors at T_c, exhibit a discontinuity between an exponential nature below (in the ordered phase) to a linear temperature dependence above (in the normal phase). This is in sharp contrast to the λ-shaped curve (representing the λ transition) for $C_v(T)$ observed near the Bose condensation temperature (T_{BC}) for He^4. The origin of the conceptual difficulty with SBB theory can be traced to the fact that in the conventional superconductors, the electron pairs are not strictly in truly bound states and that the superconducting coherence lengths are several thousand lattice spacings. At very low fermionic carrier densities, however, the above theoretical limitations are removable and the BCS theory and the SBB theory can be reconciled to some extent. This is somewhat realized in the Cuprate superconductors where the in-plane coherence lengths are only a few lattice spacings in some cases (Cava et al., 1987). Furthermore, in the field of dilute atomic gases, which are electrically charged neutral, a cross-over from BCS pairing to BEC has been observed experimentally in recent times, besides the usual BEC in some of the other cases (Ketterle et al., 1995, 2003; Cornell et al., 1995; Jin et al., 2003).

5.7   SUPERCONDUCTIVITY IN NOVEL MATERIALS

So far, our discussions have been confined to the conventional superconducting materials entirely. We now take up the cases involving other kinds of superconducting materials where the microscopic mechanism and/or the nature of pairing may differ from those described by the BCS theory with phonon mediated spin-singlet pairing producing anisotropic superconducting gap. As mentioned before, the conventional superconducting materials are mostly metals and metallic alloys. The candidates for the non-conventional and exotic superconducting materials are of wide variety. They include low dimensional organic compounds, various types of actinides and lanthanides (Kondo-Anderson systems or heavy-fermion systems), different types of doped ceramic oxide materials (high-temperature superconductors) and even some of the inter-metallics, and binary or ternary compounds involving rare earth and transition metals. In these systems, the mechanism and/or the nature of pairing symmetry are believed to often deviate very strongly from those in conventional situations. Moreover, in some of the above materials in the certain composition ranges the nature of the normal phase is very different from that expected in a metal or metallic alloy and hence cannot be characterized by the Landau Fermi liquid (LFL). This poses a challenge for the conventional scheme of Cooper pair formation as well.

In the majority of the above novel systems, magnetism, and super-conductivity exhibit a very striking interplay not seen in the conventional systems usually. We had briefly touched upon this aspect earlier in Chapters 1 and 2. This feature is an important ingredient for the formulation of the microscopic theories for superconductivity in these exotic systems. The proximity or coexistence of superconductivity with ferromagnetic and anti-ferromagnetic orderings (truly long-range or even short-range) brings into play the necessity of modifying the Cooper pairing scheme and the examina-tion of the possible nontrivial role of the magnetic excitations (magnons and paramagnons) in mediating the attractive pairing interaction. The behavior of the superconducting gap function also needs a detailed analysis. In the case of ferromagnetic spin correlations of itinerant type accompanying the onset of superconductivity, the conventional spin-singlet pairing competes with the spin-triplet pairing. Moreover, there exists a very strong possibility of the inter-electron Coulomb repulsion becoming over-screened to produce an effective inter-electron attraction because of the exchange and correlation corrections, which can even boost the singlet pairing. In the situations corre-sponding to the coexistence or proximity of itinerant anti-ferromagnetism/ spin density wave (SDW) with superconductivity, the spin-singlet pairing is quite feasible; however, the superconducting gap function may be anisotropic in the k-space. Moreover, the anti-ferromagnetic magnons/paramagnons may also mediate attractive interaction between the electrons and hence contribute substantially to the pairing mechanism. Thus, we have very strong grounds for electronic mechanism of superconductivity operating in these systems of magnetic superconductors. The materials belonging to this class are found amongst heavy fermion superconductors like $CeCuSi_2$, UPt_3, binary alloys like Y_9Co_7, doped ceramic oxides like $La_{2-x}Sr_xCuO_4$, $LaFeAsO_{1-x}F_x$, etc., and organic superconductors like $(TMTSF)_2X$, $TCNQ$, etc.

Furthermore, the normal phases of some of the ceramic oxides in the underdoped regime show strong departure from LFL behavior and therefore the standard formalism for Cooper pairing needs a major revision. Moreover, the anomalous normal state properties of these layered systems with very short coherence lengths (of the order of only a few lattice spacings in contrast to the conventional systems) in their superconducting phases, also lead to the exotic concept of "BEC of real space pairs (tightly bound fermionic pairs)" as a microscopic process for superconductivity in these systems (Chakraverty et al., 1987; Chaudhury and Chakraverty, 1999; Pal, 2017). This proposal is an alternative to the fermionic pairing theory in momentum space, as advocated by the BCS theory. It can also be looked upon as a revival and physical realization of SBB theory, in the case of the cuprate superconductors. This apart, the attractive interaction required to form the fermion pairs, in momentum-space or in real-space, is believed to have a large contribution from the non-phonon mechanisms besides the phonon driven ones. The candidates for these non-phonon channels are anti-ferromagnetic spin fluctuations, charge transfer excitons, and plasmons, in accordance with the phase diagrams of these systems. Moreover, quite interestingly, the highest T_c point in the phase diagram of these cuprates is often found to occur at the quantum critical point *viz.* at the crossover from the non-Fermi liquid phase to the Fermi-liquid phase. This aspect, unfortunately, has not been paid much attention to. There are also important puzzles like the very high magnitude of the ratio of the slope to jump in specific heat observed at T_c, which are not quite satisfactorily resolved from standard theoretical treatments. Some progress has been made in these directions though (Chaudhury, 1995; Bhattacharjee and Chaudhury, 2016; Roy Chowdhury and Chaudhury, 2018).

For the quasi-1d organic superconductors, besides the above listed non-phonon mechanisms, the side chain charge polarization can also mediate attraction between the conduction electrons moving along the main chain. This is believed to be one of the principal candidates for a microscopic mechanism for superconductivity in these systems, as was proposed by Little long ago in 1964 (Little, 1964). In some of the systems belonging to this class, the possibility of pairing mediated by charge transfer exciton seems quite plausible (Roy Chowdhury and Chaudhury, 2015) (Figures 5.1a and b).

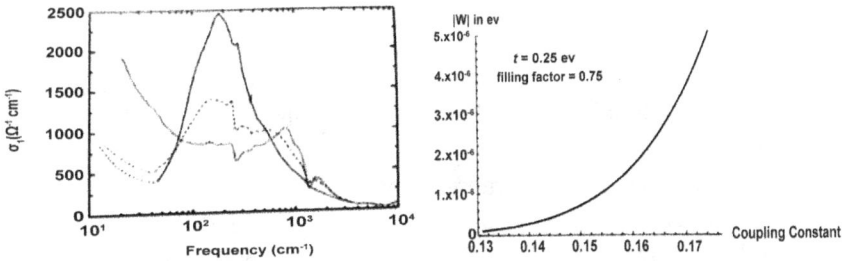

FIGURE 5.1 (a) Experimentally measured absorptivity of $(TMTSF)_2AsF_6$ vs. frequency; (b) Cooper pair binding energy vs. coupling constant in the quasi-1d model

Source: Reprinted with permission from the paper by Roy Chowdhury, S., & Chaudhury, R., (2015). *Physica B.* 2015 Elsevier).

Before concluding, it is important to look at our goals. The first and foremost target is to raise the T_c further, from the existing highest super-conducting transition temperature of 164K observed in *Hg*- based cuprates (*HgYBaCuO*) under the application of very high pressure. The dream is of course to prepare a material, exhibiting superconductivity at the room temperature. The development of such a material will be of immense help in solving the dissipation problems related to the power transmission over long distances, bringing in a technological revolution. Replacing the normal metallic wires with superconducting wires as connectors for various electronic components inside integrated circuits, would also improve upon the performances of various semiconductor-based devices tremendously. For these futuristic developments, a close collaboration between the condensed matter physicists and the material scientists is essential. As pointed out earlier, within the framework of the pairing theory the frequency of the boson mediating the pairing interaction and the attractive coupling constant will have to be adjusted optimally, so as to get a higher T_c. This requires a detailed study and an understanding of the microscopic processes as well as the material structures and a proper symbiosis between the two approaches. Very often, this would also need the use of multi-layer and multi-band generalization of the microscopic pairing theory (Chaudhury, 2009; Chaudhury and Das, 2020). Besides, technological utility and more efficient performances of the devices built from them would have to be kept in focus too.

The other aims are more academic in nature. The investigation of the interplay between superconductivity and various kinds of charge orderings and magnetic orderings in the same material deepens our understanding of the many-body physics and the statistical mechanics particularly for low dimensions. Therefore, various striking features observed in the novel and exotic materials can play very important roles in providing new insights into the fundamentals of condensed matter physics as well as giving possible clues to the attainment of higher T_c.

There is also a symbiosis emerging between the pure academic or basic research and technological research. This includes the problems involving various low-dimensional superconductors like cuprate superconductors and organic superconductors, superconductor-normal junctions, exotic topological insulator-superconductor junctions, topological superconductors, etc. These also have a large overlap with certain challenging problems in quantum field theory and even high energy physics besides enlarging the concepts of superconducting pairing theory itself under various normal state conditions and scenarios. The advanced material research and innovation of new devices for electrical transport as well as magnetic measurements are quite likely to benefit from these investigations too.

We close this chapter on a very happy note and with optimism. Very recently, i.e., in 2015 a new purely three-dimensional chemical compound belonging to the Hydride family has been synthesized, which exhibits T_c as high as 203K under a very high applied pressure of magnitude of about 100–200 Giga Pascal (Drozdov et al., 2015). The necessity of the application of such huge pressure is to cause an insulator-metal transition in the normal phase in the first place, it seems. The compound is most likely to be of the stoichiometric composition H_3S (Drozdov et al., 2015). At first glance, the mechanism for superconductivity in this system seems to be the conventional one *viz.* phonon-based and the very large magnitude of T_c is probably due to the very high value of the Debye frequency. This is in tune with the theoretical conjectures and suggestions of V. L. Ginzburg (with others) and N. W. Ashcroft, put forward independently much earlier, regarding the possibilities of obtaining very high superconducting transition temperatures in metallic Hydrogen and its compounds under the application of huge external pressure (Ginzburg and Kirzhnits, 1982; Maksimov and Savrasov, 2001; Ashcroft, 2004). Later in another class of similar systems *viz.* Lanthanum hydride and superhydrides and carbonaceous sulfur hydride system, application of very high pressure has led to a

T_c of 250K and above (nearing 290K) (Drozodov et al., 2019; Somayazulu et al., 2019; Snider et al., 2020). Besides, T_c as high as 236K has been reported from experiments performed on a nanostructured $Ag - Au$ binary system (made of two noble metals!) under ambient conditions (Thapa and Pandey, 2018). More experimental analysis however is required to have the confirmation and the full understanding of the occurrences of superconductivity in these new systems. Nevertheless, this series of latest discoveries have shown a lot of realistic promise towards achieving our dream of obtaining room temperature superconductivity (Figure 5.2).

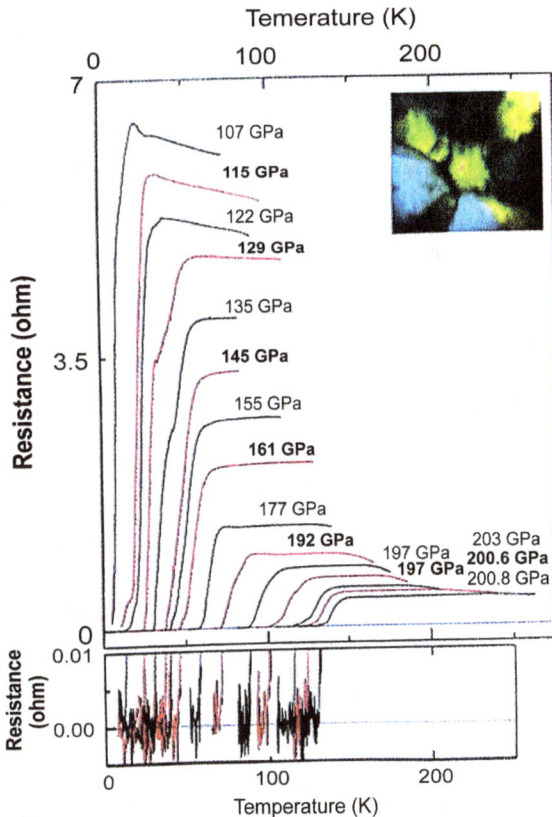

FIGURE 5.2   Resistivity vs. temperature under various magnitudes of external pressure for H_3S.

Source: Reprinted with permission from the paper by Drozdov et al., (2015). *Nature.* 2015 Springer Nature).

KEYWORDS

- Bardeen-Cooper-Schrieffer
- Bose-Einstein condensation
- Gorter-Casimir
- Scafroth-Blatt-Butler
- superconducting quantum interference device
- superconductor-normal metal

REFERENCES

Abrikosov, A. A., Gorkov, L. P., & Dzyaloshinskii, I. E., (1975). *Methods of Quantum Field Theory in Statistical Physics*. Dover Publications, New York.

Anderson, M. H., Ensher, J. R., Matthews, M. R., Wieman, C. E., & Cornell, E. A., (1995). *Science, 269*, 198.

Ashcroft, N. W., & Mermin, N. D., (2017). *Solid State Physics*. Brooks Cole/Cengage Learning India.

Ashcroft, N. W., (2004). *Phys. Rev. Lett., 92*, 187002.

Aynajian, P., Keller, T., Boeri, L., Shapiro, S. M., Habicht, K., & Keimer, B., (2008). *Science, 319*, 1509.

Bhattacharjee, S., & Chaudhury, R., (2016). *Physica B., 500*, 133.

Blatt, J. M., (1964). *Theory of Superconductivity*. Academic Press, Michigan.

Brandt, E. H., & Das, M. P., (2010). *Cond-mat arXiv, 1007*, 1107.

Carbotte, J. P., (1990). *Review of Modern Physics, 62*(4), 1027.

Cava, R. J., et al. *Phys. Rev. Lett., 58*, 1676.

Chakraverty, B. K., Feinberg, D., Hang, Z., & Avignon, M., (1987). *Solid State Commun., 64, 8*, 1147.

Chaudhury, R., & Carbotte, J. P., (1993). *Solid State Communications, 85*(12), 1039.

Chaudhury, R., & Chakraverty, B. K., (2000). *Criterion for Bound State Formation in Layered Materials*. SNBS-CNRS Research Report.

Chaudhury, R., & Das, M. P. (To be communicated).

Chaudhury, R., & Das, M. P., (2013). *Jour. Phys. Condens. Mat., 25*, 12202.

Chaudhury, R., & Jha, S. S., (1984). *Pramana Journal of Physics, 22*, 431.

Chaudhury, R., (1995). *Canadian Journal of Physics, 73*(7–8), 497.

Chaudhury, R., (2009). arXiv: 0901.1438v1 [cond-mat.supr-con].

Davis, K. B., Mewes, M. O., Andrews, M. R., Van Druten, N. J., Durfee, D. S., Kurn, D. M., & Ketterle, W., (1995). *Phys. Rev. Lett., 75*(22), 3969.

Drozdov, A. P., Eremets, M. I., Troyan, I. A., Ksenofontov, V., & Shylin, S. I., (2015). *Nature, 525*, 73–76.

Drozdov, A. P., Kong, P. P., Minkov, V. S., Besedin, et al., (2019). *Nature, 569*, 528.

Delft, D. V., (2007). Freezing Physics – Heike Kamerlingh Onnes and the quest for cold. Royal Netherlands Academy of Arts and Sciences.

Eliashberg, G. M., (1960). *Sov. Phys. JETP, 11,* 696 and *Zh. Eksp. Teor. Fiz., 38,* 966 (1960).

Frohlich, H., (1950). *Proc. of the Phys. Soc., 63*(7), 778.

Ginzburg, V. L., & Kirzhnits, D. A., (1982). *High Temperature Superconductivity.* New York: Consultants Bureau.

Greiner, M., Regal, C. A., & Jin, D. S., (2003). *Nature, 426,* 537.

Little, W. A., (1964). *Phys. Rev., 134,* A1416.

Mahan, G. D.,(1981). *Many Particle Physics.* Plenum Press.

Maksimov, E. D., & Savrasov, Y, D., (2001). *Solid State Communications, 119,* 569.

McMillan, W. L., (1968). *Phys. Rev., 167,* 331.

Moschalkov, V., Menghini, M., Nishio, T., Chen, Q. H., Silhanek, A. V., Dao, V. H., Chibotaru, L. F., Zhigadlo, N. D., & Karpinski, J. (2009). *Phys. Rev. Lett., 102,* 117001.

Onnes, H. K., (1911). *Akad. van Wetenschappen, 14*(113), 818.

Pal, S. D., (2017). *Jour. Phys. Commun.,* doi: 10.1088/2399-6528/aa9398.

Pines, D., (1999). *Elementary Excitations in Solids.* Advanced Books Classics, Perseus' Frontiers in Physics Series and Westview Press.

Plakida, N. M., (1995). *High-Temperature Superconductivity: Experiment and Theory.* Springer Verlag.

Roy, C. S., & Chaudhury, R., (2018). *arXiv: 1807.11188v1 [Cond-mat.supr-con].*

Roychowdhury, S., & Chaudhury, R., (2015). *Physica B., 465,* 60.

Schrieffer, J. R., (1983). *Theory of Superconductivity.* Advanced Books Classics, Dover.

Snider, E., Gammon, N. D., Mcbride, R., Debessai, M., Vindana, H., Vencataswamy, K., Lawler, K. V., Salamat, A., & Dias, R. P., (2020). https://doi.org/10.1038/s41586-020-2801-z.

Somayazulu, M., Ahart, M., Mishra, A. K., Geballe, Z. M., Baldini, M., Meng, Y., Struzhkin, V. V., & Hemley, R. J., (2019). *Phys. Rev. Lett., 122,* 027001.

Steglich, F., Bredl, C. D., Lieke, W., Rauchschwalbe, U., & Sparn, G., (1984). *Physica B., 126,* 8291.

Thapa, D. K., & Pandey, A., (2018). *Evidence for Superconductivity at Ambient Temperature and Pressure in Nanostructures.* Private Communications (communicated to journal).

Zwierlein, M. W., Stan, C. A., Schunck, C. H., Raupach, S. M. F., Gupta, S., Hadzibabic, Z., & Ketterle, W., (2003). *Phys. Rev. Lett., 91,* 250401.

CHAPTER 6

Interface of Molecular Biophysics with Condensed Matter Physics

6.1 INTRODUCTION

The two emerging and major research fields *viz.* molecular biophysics and condensed matter physics are apparently very remotely connected. At best, some interface between soft matter physics or statistical physics-based modeling and certain bio-molecular (like DNA, RNA, or Protein related) properties have been developed in recent times with some amount of success. Often, however, the underlying vital molecular biophysical and biochemical processes of DNA carrying the substantial signature of quantum effects, are not included in these modelings. In fact, only a few macroscopic physical and chemical properties are usually dealt with in this interdisciplinary approach. On the other hand, quantum chemistry based ab-initio approaches have been employed to study the structure and properties of various biomolecules. The molecular-level processes include mutation of DNA, replication of DNA, and transcription of DNA (to generate RNA), Protein synthesis, and its conformational dynamics as examples. Very interestingly, many of these processes share a lot in common/ parallelism with the microscopic features of various condensed matter phenomena. In particular, it turns out that the structure and dynamics of bio-molecules like DNA can be modeled quite well by processes similar to ones encountered in the field of quantum magnetism or quantum transport corresponding to low-dimensional systems. This takes one to substantially deep inside the domain of quantum physics. As a result, various concepts and techniques used in theoretical condensed matter physics become readily available and extremely useful in theoretical molecular biophysics. Thus, one can deal with problems involving tautomeric mutation (TM) in DNA, melting of DNA, and microwave (MW) effect in DNA quite systematically by making use of these resources borrowed from theoretical condensed matter physics. Besides, at a slightly softer level, the spatial distribution of nucleotides in DNA and the property of segmentation can be studied very satisfactorily with the help of techniques often used in statistical mechanics. Again, many of these methodologies have considerable overlap with certain aspects of condensed matter physics as well. This kind of novel interdisciplinary approach has enriched a large area of theoretical molecular bio-physics in recent times and has helped to understand various biophysical processes from certain fundamental principles of quantum physics and more precisely condensed matter physics. This endeavor has in fact led to a new field named "quantum biology."

Very recently *viz.* in 2016, the award of Nobel Prizes in Medicine and Chemistry for contributions in the areas of recycling processes exhibited by cells inside living organisms and the designing of molecular machines respectively, have undoubtedly provided a new impetus to the research in quantum biology.

In this chapter, we will mainly focus on the following areas:

1. TM of DNA and RNA and their repair via quantum modeling; melting of DNA.
2. Effect of TM on protein synthesis.
3. Study of segmentation of DNA and the role of statistical physics in locating the exon-intron boundary. We also briefly discuss the characterization of the spatial distribution of the nucleotides in a DNA strand belonging to different species, from the perspective of statistical physics.
4. Study and modeling of the effect of MW irradiation on DNA and other biological systems from quantum condensed matter physics viewpoint.
5. Review and application of our approach to the phenomenon of DNA melting.

The double-helix structure of the DNA molecule is well known and can be seen in many textbooks (e.g., Griffiths et al. (2002). Chapter 2, p. 28). A DNA molecule consists of 2 helices inter-twinned and is made up of "nucleotides" which serve as 'building blocks.' The nucleotides in turn consist of sugar-phosphate backbones as well as bases. The chemical bonding between the complementary bases on the two strands is based on hydrogen bonds, whereas that between a sugar and a phosphate is known as phosphodiester bond (Griffiths et al., 2002).

6.2 TAUTOMERIC MUTATION (TM) THROUGH QUANTUM MODELING

The mutation is a very important biological process for all living organisms whether macro or micro. It has tremendous pathological consequences for medical science. Mutation is believed to play a very crucial role in the evolution and the appearance of the large variety of species on this earth in accordance with Darwinian Theory. The deviation from the usual base-pairing rule (BPR) of Watson and Crick for a double-stranded DNA (deoxyribonucleic acid) during the replication process is known as mutation. This can be triggered either by electromagnetic irradiation and chemical contamination or by quantum fluctuations. The last one is the source of TM. To elaborate slightly, all four bases (nucleotides) in a DNA molecule, *viz.* A (Adenine), T (Thymine), C (Cytosine), and G (Guanine) can exist in two different states. These states are known as "keto" and "enol." In fact, the bases undergo constant fluctuations between these two states. In the keto state, the base pairs obey the usual BPR, i.e., 'A' is paired with 'T' and 'C' is paired with 'G.' In contrast, in the enol state, the pairings can be quite anomalous *viz.* 'A' may be paired with 'G' and 'C' with 'T' for instance. These conjugate base pairs are connected by hydrogen bonds. Now hydrogen atom being the lightest one, it undergoes quantum hopping with a very high amplitude. This process, which can also be looked upon as a proton tunneling, alters the hydrogen bond arrangements within the bases themselves. This in turn leads to the creation of tautomers (very much like isomer) of the original bases involved in pairing connecting the two strands. Again, these tautomers look very similar chemically to the enol states of other (non-conjugate) bases and this striking feature leads eventually to anomalous base pairings in a DNA molecule during replication. This is briefly the origin of TM.

Very interestingly, however, this very quantum process which brings about mutation can also cause restoration of original base pairing in a scheme known as "complementarity restoration (CR)," as we will explain later.

It may further be noted that the nucleotides (bases) situated adjacent to each other on an individual strand, have covalent bonding between them.

Moreover, the strength of this covalent bond is much higher (at least 10 times) than that of the hydrogen bond involved in the pairing of the bases on the two complementary strands. The phenomena of TM and the CR in DNA through successive replication processes have been schematically shown in the figures in Griffiths et al. (2002) (Chapter 10, p. 320) and the reader may have a look at it.

The above facts combined with the quantum origin of TM lead to the following possible quantum spin modeling of this process (Chaudhury, 2007).

We view a double-stranded DNA molecule as a 2-legged Ising spin-isospin ladder system in the presence of a generalized transverse external field. Furthermore, the external field couples to both transverse components of spin and iso-spin degrees of freedom. According to this model, the nucleotides 'A,' 'T,' 'C,' and 'G' are regarded as different quantum states of a DNA molecule at any local site within the molecule. They are represented as $\left(X, \frac{+1}{2}\right), \left(X, \frac{-1}{2}\right), \left(Y, \frac{+1}{2}\right), \left(Y, \frac{-1}{2}\right)$ respectively, where the first letter provides the iso-spin label and the second number stands for the spin state ("up" or "down"). Again, 'X' and 'Y' themselves are called "iso-spin upstate" and "iso-spin downstate" respectively for convenience. In this representation, the inter-strand coupling between the iso-spin operators is of ferromagnetic type whereas, that between the longitudinal components (z-components) of the spin operators is of anti-ferromagnetic nature. It is also worth pointing out that the transverse field couples only to the intra-strand degrees of freedom or operators, for both spin as well as iso-spin sectors. This transverse coupling part of the total Hamiltonian has terms which induce flips for spin and iso-spin states, mimicking mutations and thereby lead to violation of BPR. The quantum nature of this model becomes very explicit in the presence of this transverse external field. The same modeling can be done for a RNA molecule as well, with T replaced by U.

Mathematically our above quantum spin-iso-spin model, to be called simply as "quantum spin model" from now onwards, can be expressed as:

$$H_{QSM} = H_{\text{I sing}}^{\text{inter-strand}} + H_{\text{T ransverse}}^{\text{intra-strand}} \qquad (6.1)$$

where, two parts are given by:

$$H_{I\,sing}^{inter\text{-}strand} = \sum_{i,M,N,s,s'} \lambda_i^{1,2} \sigma_{i,M,s;1}^{z} \sigma_{i,N,s';2}^{z} \delta_{M,N} \delta_{s,-s'} \tag{6.2}$$

and

$$H_{Transverse}^{intra\text{-}strand} = \sum_{i,l} \frac{1}{2} H_i' \left[\left(\sigma_{i,l}^{+} + \sigma_{i,l}^{-}\right)_{sp} XI_{psp} + I_{sp} X \left(\sigma_{i,l}^{+} + \sigma_{i,l}^{-}\right)_{psp} + \frac{1}{2} \left(\sigma_{i,l}^{+} + \sigma_{i,l}^{-}\right)_{sp} X \left(\sigma_{i,l}^{+} + \sigma_{i,l}^{-}\right)_{psp} \right]$$

$$\tag{6.3}$$

It should be emphasized here that the operators σ's occurring in the above equations are in fact composite operators comprising of both spin operators and iso-spin operators; the symbol I stands for the identity operator. The parameter H denotes the transverse external field which we have assumed to be spatially uniform, i.e., site-independent, for simplicity in our calculations and $\lambda_i^{1,2}$ denotes the inter-strand coupling involving the corresponding conjugate sites.

Before we go over to the other type of quantum modeling based on the bosonic approach, we would like to first explore the consequences of the above spin modeling itself and its importance for the mutational processes involving DNA.

In the situation when the strength of the inter-strand Ising coupling is much larger than that of the intra-strand transverse coupling, one can treat the second part of the Hamiltonian perturbatively and use the ground state of the first part as a reference state. This state obviously satisfies BPR. Assuming the coupling strength involving the bonds $A-T$, and $G-C$ to be the same, the unperturbed ground-state turns out to have a degree of degeneracy 4^N, where N is the total number of nucleotides along any strand. Employing degenerate perturbation theory in combination with Fermi Golden Rule, detailed calculations lead to the following expressions corresponding to the rate of TM and the rate of CR, respectively, assuming a very large N for a macroscopic size of DNA molecule:

$$\tau_{(NBPR)}^{(-1)} = \frac{(2\pi)^2}{h} \sum_{i,f} \rho_i \left| F1(i, f = ex) \right|^2 \delta(\omega_{fi}) \tag{6.4}$$

and the equation for $\tau_{(BPR)}^{(-1)}$ is exactly the same as the above one with only F_1 replaced by F_2, where the quantities F_1 and F_2 are given by:

$$F1(i, ex) = \left\langle \psi_i \left| H_{Trans}^{IS} \right| \psi_{ex} \right\rangle \tag{6.5}$$

and

$$F2(i,j) = \sum_{ex} \left\langle \psi_i \left| H_{Trans}^{IS} \right| \psi_{ex} \right\rangle \left\langle \psi_{ex} \left| H_{Trans}^{IS} \right| \psi_{ej} \right\rangle \tag{6.6}$$

where, $|\psi_i\rangle$ is a 'parent state,' $|\psi_{ex}\rangle$ is a 'daughter product state' (product of mutation after one round of replication) and $|\psi_j\rangle$ is a 'granddaughter product state' (generated after two rounds of mutation). Again ρ_i is the probability of occupation of the initial chosen ground state denoted by index i, which is often independent of i and is equal to 4^{-N}. The other quantity ω_{fi} is the equivalent frequency corresponding to the energy difference ΔE between the ground state represented by i and the final (lowest excited) state denoted by f. It may be pointed out that the 'daughter state' does not obey BPR; however amongst the 'granddaughter states' many of them are consistent with BPR and thus belong to the degenerate manifold of 4^N states, as the 'parent state' does.

From the above set of equations, one can calculate the repairing efficiency (RE), defined as the ratio of the rates of CR and TM. The magnitude for RE turns out to be $2H^2$, H being the uniform transverse external field. This estimate of RE gives a rate of compensation of the TM s by the second-order processes involving the quantum transverse coupling. This rate can then be compared with the experimental rate of the corresponding process for the real DNA of a living system. Fortunately, there are classes of bacteria for whom the data is available. In this comparison scheme, the external field H is treated as a phenomenological parameter.

It is worthwhile to point out that a real DNA molecule, even at zero temperature, has an intrinsic coupling to the environment through direct electromagnetic interaction or chemical interaction. This causes "quantum decoherence" of the parental state if it is a pure state. It can be shown that this would lead to the RE to be much less than $2H^2$. Besides, in a real DNA system, both quantum decoherence and the thermal mixing of the parental states may have a considerable contribution from the nontautomeric mutational processes as well.

Coming back to the possible application of our calculational results to a real biological system, a bacterial organism (fungus) named "Neurospora" turns to be experimentally quite well studied. We choose this very species for our phenomenological treatment. It is observed from experiments that in the absence of any mutagenic treatment, the effective spontaneous mutation rate (after the primary repair processes or so-called "proofreading") is about 2×10^{-11} per second per nucleotide for the above micro-organism. Assuming the number of mutation channels, i.e., mutated daughter states allowed after the 1st round of replication for this species to be n, we arrive at the following equation,

$$H^2 n = 2N10^{-11} \qquad (6.7)$$

where, we recall that in fact $N = 2n$. This gives the approximate magnitude of H as 3.10^{-6} in some appropriate unit. The above equation also yields the magnitude of the secondary RE to be of the order of 10^{-11}. Therefore, the secondary repairing process driven by quantum fluctuations can compensate about 2 out of every 10^{11} point mutations in Neurospora. The detailed investigation however reveals that the ratio of this secondary repair rate and the intrinsic enzymatic action based primary repair rate is generally of the order of 1% only. Nevertheless, this quantum mechanical perturbative effect manifested in the emergence of the secondary repair rate, points very clearly to the role of condensed matter physics-based modeling in making qualitative and quantitative predictions for molecular biophysical processes. It would however be interesting and very useful if experimental results for DNA mutation are available from more such micro-organisms.

Later, we have taken into consideration the asymmetry in the magnitude of the coupling constants involving the G-C bond and the A-T bond and carried out a more realistic analysis for the degeneracy factor and the RE (Chaudhury et al., 2019).

An important aspect of this condensed matter physics-based treatment of the quantum mechanically driven TM problem is the existence of a parallel theoretical proposal providing an alternative description of this phenomenon, in terms of a bosonic model (Golo and Volkov, 2003). The Hamiltonian in this case is an intra-strand one and consists of two types of contributions viz. site energy term and inter-site hopping term for the bosons. These bosons are essentially 'excitonic' in nature in the sense that they represent single site tautomeric mutational excitation in any strand of the DNA. This model is based on the assumption of tunneling states of protons as two-level systems. Mathematically, it is expressed as:

$$H_{bosonic} = \sum_n E_0 b_n^+ b_n + k \sum_n \left(b_n^+ b_{n+1} + b_{n+1}^+ b_n \right) \qquad (6.8)$$

where, the operators b_n and b_n^{+} represent the bosonic destruction and bosonic creation operators respectively at the nth site on a strand. The quantity E_0 is the bosonic site energy and is in fact the energy of the exciton. The second term describes the hopping of this boson between the nearest neighbor sites on any one strand, indicating the intrastrand propagation of mutation with an amplitude κ. This hopping amplitude depends upon the dipolar interactions between the nearest neighbor bases apart from many other factors. The above Hamiltonian is analogous to the very well known tight-binding model corresponding to the band theory of fermions in condensed matter physics.

It is very interesting to explore the connection between these two types of quantum modeling described above *viz.* spin modeling and bosonic modeling. We now discuss in detail the physical and mathematical relationship between them and provide a short derivation on this aspect.

First of all, in the bosonic approach the proton tunneling process combined with the hopping of the excitation lead to a deformation of elastic DNA through the π-electron redistribution. In the quantum spin model perspective on the other hand, the transverse field H can be linked to the physical relocation and reorientation of the Hydrogen bonds. This causes phonon-like distortion modes to be created due to the charge redistribution, along the 'rungs' joining the multiple strands generated in the course of the replication process. This aspect can be expressed mathematically as:

$$H_q = C_q (v_q + v_{-q}^{+})$$
(6.9)

where, v_q and v_q^{+} are the phonon destruction and creation operators respectively and the coefficients C_q contain the proton tunneling matrix elements.

Thus the existence of an intrinsic coupling between the electronic and ionic degrees of freedom and of the deformation modes in both the models, provide a very important conceptual bridge between the two approaches.

For building the connection between the two models, the following assumptions and methodologies are followed:

1. It is assumed that both the quantum models become operational physically only after the completion of many rounds of replication,

enabling the formation of a large number of mutually complementary strands in the system in accordance with the BPR.

2. There exist long-range orderings for both spins and iso-spins with the manifestation of anti-ferromagnetic ordering for the former and the ferromagnetic ordering for the latter, along the rungs. The arrangements of spins and iso-spins along the strands are however assumed to be totally random.

3. The TM implies excitations in either or both spin channels and isospin channels, brought about by the action of the transverse field H.

4. The quantum spin-iso-spin Hamiltonian is bosonized by Holstein-Primakoff transformations (HPT) for both spin and iso-spin sectors. It is worth recalling that HPT has been dealt with in great detail in Chapter 2.

5. Treating the inter-strand term as the principal term and the intra-strand one as a perturbation in the bosonized version of the spin Hamiltonian, Schrieffer-Wolf transformation (SWT) is carried out. Thereafter terms up to 3rd order in the intra-strand contribution are retained. It may be recalled that SWT has been discussed in some details in Chapter 2 earlier.

This new effective bosonized Hamiltonian so generated, looks very similar to the proposed bosonic model Hamiltonian. However, this effective model contains the hopping of two distinct kinds of bosons in contrast to the presence of only one kind of boson in the originally proposed bosonic model. Besides, this newly generated model contains an additional contribution of 2nd order in the intra-strand part as well, which has no analog in the proposed bosonic model. These aspects will become transparent after the calculations. We now very briefly present the calculations below:

Recalling that in the quantum spin model, the principal part is identified with H_{Ising} and the perturbation part with $H_{Transverse}$. The standard SWT then generates the following effective Hamiltonian:

$$H_{eff} = H_{1\,sing} + H_{Transverse} + \left[H_{QSM}, S \right] + \left[\left(\frac{1}{2} \right) \left[\left[H_{QSM.S} \right] S \right], S \right] + higher order terms (neglected)$$

$$(6.10)$$

We have retained terms up to 3rd order in perturbation in the right-hand side of the above equation, as stated earlier. The operator S can be determined from the following equations for its matrix elements,

$$S^{mn} = \frac{H^{mn}_{Transverse}}{E_n - E_m} \tag{6.11}$$

where, the corresponding matrix elements of the operators have been related with E's being the eigenvalues of H_{Ising}, which are $\frac{-\lambda_i}{4}$ and zero only, as is very clear from its form.

Making use of the above set of equations, it is possible to extract the corresponding matrix elements of the operator term which is of 3rd order in $H_{Transverse}$ and which occurs in H_{eff}. Considering the two strands, we need to evaluate totally 16 matrix elements. With this, one can write down the full formal mathematical expression for H_{eff} which would be very long.

Next, we show that the above 3rd order term in $H_{Transverse}$ indeed contains the hopping term of the proposed bosonic model and that the H_{Ising} of the quantum spin model can be identified with the site energy term of the bosonic model as well. For this, we carry out the bosonization operation based on HPT corresponding to both spin and iso-spin sectors as discussed earlier.

Avoiding the presentation of very long mathematical calculations, details of which can be found in the paper published in NMFM (Chaudhury, 2015) [cited in the reference section at the end], we only display the final result viz.

$$(E_0)_{bosonic} = \left(\frac{1}{8}\right)\lambda^{1,2} \tag{6.12}$$

and

$$k \sim \left(\frac{1}{4}\right)H^3 \tag{6.13}$$

Thus, we have been finally able to identify the parameters occurring in the bosonic model with those of the quantum spin model, as exhibited in the equations above. This, therefore, implies a unification of the two radically different approaches based on quantum modeling. It must however be emphasized, that more correct or exact relations between the coefficients or parameters of the above two models can be obtained only after handling the full and very long-expression for H_{eff}.

We now make a few concluding remarks on the two proposed quantum models discussed in this section:

1. The unification scheme based on the low order perturbation expansion coupled with bosonization of the quantum spin model brings out the nature of the bosons occurring in the bosonic model more precisely and in a very clear physical way. Besides, these bosons are found to be of two distinct types, arising from the spin and the iso-spin sectors.

2. The phonons representing the deformation of elastic DNA are implicitly present in these calculations through the presence of a transverse field in the quantum spin model and proton tunneling in the bosonic model. This also implies a hidden coupling between the phonons and the magnons/pseudo-magnons, which is quite important for many other phenomena.

3. The lowest order processes present in the calculations here, involve 1 phonon-1 magnon/1 pseudo-magnon and 1 phonon-3 magnon/3 pseudo-magnon couplings. For our unification calculations, however, we have only retained the 1 phonon-1 magnon/ 1 pseudomagnon process in the very lowest order for simplicity.

4. In the quantum spin model description, each strand has been assumed to be of paramagnetic nature with respect to both spin and isospin degrees of freedom. In reality, the presence of covalent bonding leads to a correlation between the above degrees of freedom with a slightly different expression for the RE. The unification relation between the quantum spin model and the bosonic model derived here remains unaffected though.

5. It is known from chemical analysis that the bond strength or the affinity involving the $G–C$ bond is higher than that of the $A–T$ bond in reality. Incorporation of this feature would modify our originally proposed spin model by inclusion of two different coupling constants for the inter-chain Ising interactions. The unification algebra would also have to be redone in that case.

6. It is an experimentally verified fact that real DNA mutations in living organisms are governed by 'adaptive mutations' (Cairns et al., 1988). To slightly elaborate on this, the probability of

mutation at all the DNA sites is not the same; rather it depends upon the local chemical environment in the related corresponding chromosome. Therefore, more rigorous calculations based on quantum modeling in the future must take care of this fact as well.

6.3   EFFECT OF TAUTOMERIC MUTATION (TM) ON PROTEIN SYNTHESIS

First of all, it is important to note that the cross-sectional areas of both DNA and Protein being in the nanoscale regime *viz.* 2 nm and 20 nm respectively, the single-particle quantum tunneling effects can be of special importance there (McFadden, 2000). As a result, the tautomeric fluctuations in DNA are expected to vary strongly affect the Protein synthesis as well as the Protein properties. The simplest source of this is through the drastic change in the type of amino acids generated, induced by the mutations in the parental DNA during 'transcription' to the RNA. This then in turn through the process of 'translation' with the formation of peptide linkages of the amino acids, can lead to unwanted protein formation. As an illustration, the 'triplet codon' (or simply 'codon') *GCC* codes for amino acid Alanine whereas, the codon *GGC* codes for Glycine. Thus, just a point mutation in DNA during replication or transcription can eventually cause a drastic change in the nature of amino acids produced. In some cases, this undesirable amino acid production may even lead to deadly diseases. For example, the mutation-induced change in just one amino acid involved in the formation of one of the Globin proteins of blood Hemoglobin, leads to the human genetic disease Thalassemia (McFadden, 2000). A ray of hope in partially recovering from or in the reduction of such unwanted amino acid production rests on the feasibility of the CR process in DNA, which helps in undoing the mutations. It is driven by higher-order mutational effects through the quantum channels, as explained in detail in an earlier section (see the section on TM). Developing a suitable quantum/quantum chemical modeling of the sequence of biophysical processes culminating in the generation of proteins, would certainly help in displaying the consequences of the mutational effects in DNA on protein synthesis in a more transparent way.

6.4   SEGMENTATION OF DNA

The genome of any organism is made up of DNA. It however contains many "useful" and "useless" components. The useful components are called the 'genes' which take part in protein synthesis via the processes of 'transcription' and 'translation.' These involve the generation of m-RNA and subsequent production of amino acids through codons. For eukaryotes (the multicellular organisms like plants, animals, and fungi), the genes are further subdivided into "exons" (active regions or coding regions) and "introns" (inactive regions or non-coding regions) (Griffiths et al., 2002). In between various genes within a genome, lie non-genic regions called "flanks." The flanks are the so-called useless components, apparently; although they are not so well studied or understood yet.

One of the very interesting and challenging problems in molecular biophysics is the identification of the location of the border between the exons and introns, both experimentally and theoretically. A combination of the techniques involving bioinformatics and statistical physics seems to be quite powerful to address this issue (Karlin, 2005). Below, we very briefly sketch out one such procedure.

As a first step, the spatial distribution of the nucleotides is studied for the genomes corresponding to various species belonging to the class of eukaryotes (Som et al., 2003). Different well known statistical distribution functions like Gaussian, Poissonian, and Hypergeometric are tried out. An interesting result emerged that the $ln(\sigma)$ versus $ln(l)$ plot where σ denotes the variance and l represents the size of the finite sequence, known as the 'window size,' deviates from linearity behavior quite strongly (Som et al., 2003).

This implies that the scaling exponent α defined as $\dfrac{d\sigma}{dl}$ is itself l dependent. This was found to be true for both eukaryotes and prokaryotes (unicellular organisms like bacteria) with the genomic sequence analysis done for both of them. Furthermore, the departure of the magnitude of α from 0.5 indicates the presence of correlations (non-randomness) in the nucleotide distributions. Detailed statistical treatments involving power spectrum $S(f)$ could also throw light on the segmentation properties of DNA with the emergence of critical window sizes l_c corresponding to an abrupt change in the value of α. Therefore, the existence of this window size is a genuine manifestation of the segmentation property. Thus, this theoretical/computational procedure indicates the realistic existence of the phenomenon of segmentation and provides an estimate for the possible spatial location of the boundaries separating exons and introns as well.

The other notable alternative theoretical and computational approaches for the determination of segmentation in DNA are based on 'random walk model' (RWM), 'Jensen-Shannon divergence criterion' (JSDC) and Karlin et al. proposed 'statistical scoring scheme' (SSS). They are briefly discussed below.

In the RWM approach, the nucleotide distribution is characterized by the distribution of the 'minimum distances' to be covered in order to go from any position of any given type of specific nucleotide (say '*A*') to the nearest location of that very type of nucleotide in a given DNA strand. In the process, the representative point executes a 'random walk' over the other type (non-*A*) of nucleotides. This scheme is analogous to the trajectory mapped out by a 'random walker' with 'waiting time' or 'pausing time' at a position, as is often dealt with in statistical physics problems on diffusion and transport in real systems with the disorder (Zaburdaev, 2006). The so-called 'waiting time' (equivalent or proportional to the minimum distances described above in the DNA problem) distribution studied over various length scales within the genome, gives a quantitative estimate of the degree of segmentation of the DNA.

In the JSDC approach, on the other hand, an information-theoretic scheme is implemented (Lin, 1991). This approach finds applications to the problems in the wide areas like signal analysis, robotics, medical diagnostics, and linguistics as well (Katabeh et al., 2015). The entire genomic sequence is first partitioned into 'artificial fragments' in various arbitrary ways. Then the 'Shannon entropy' of the resulting configuration and thereby the 'Jensen-Shannon divergence function (JSDF)' are calculated by using the standard expressions in each case (Shannon, 1948). The specific multi-partitioning arrangement which produces 'maximum dissimilarity/ divergence' between the fragments created, manifested as a maximum of the JSDF, is considered as that representing the real biological segmentation of the genomic sequence, after an iterative procedure is carried out involving the partitioning of the entire genome in stages.

The 'SSS' methodology for defining biological segmentation of DNA is based upon the principle of chemical affinity amongst the nucleotides, as proposed by Karlin and several other researchers (Karlin et al., 1989; Braun and Muller, 1998). This in general, can include the chemical contributions from both intra-strand covalent bonds and interstrand hydrogen bonds. A probability distribution generated from the relative frequencies of occurrences of the nucleotides themselves or their 'classes' (based on some chemical or physical criterion) in a DNA strand corresponding to a given organism, is studied for various patches (spatial domains) within the genome. A sharp change in the probability distribution in terms of statistical significance, when moving from one patch to the other, implies segmentation. Application of the above principle with detailed analysis leads to the development of a "SSS" indicating and representing relative abundance of the nucleotides or their classes in various parts or patches of the DNA. On the basis of the magnitude of this score, a quantitative criterion is set for the location of the segmentation boundaries theoretically.

6.5 MICROWAVE (MW) IRRADIATION ON DNA FROM VARIOUS ORGANISMS AND ALSO ON VIRUSES AND BACTERIA

This section deals with both theoretical and experimental investigations. The theoretical studies highlight the role of thermal as well as non-thermal *viz.* resonant absorption contributions for the MW-DNA interaction. For the experimental aspects, various investigations involving the observed irradiation effects on the DNA in the brains of mice, sterilization of toothbrushes directly through MW irradiation carried out by dentists involving the DNA from viruses and bacteria and also the effects of radiation emitted from cellular phones as well as from MW ovens on cells of different types of living organisms including humans, are highlighted (Woolard et al., 2002; Golo, 2005; Demidova et al., 2012; Gujjari et al., 2011; Deshmukh et al., 2013; Banik et al., 2003).

The effect of MW radiation in the Giga and tera-hertz range on the biological systems has been of considerable research interest during the last several years (Woolard et al., 2002). It is generally believed that the MW irradiation has an important effect on the macromolecules like DNA and proteins. Nevertheless, the specific mechanism involved still requires considerable understanding, both theoretically and experimentally. In particular, the interaction of MW with biological systems can be due to the thermal effects produced by the irradiation or due to the resonance interaction of the MW radiation with the dipoles present in the molecules or both. This non-thermal or resonance contribution includes both electric and magnetic effects. This investigation is quite important because it can throw a lot of light on the mutation mechanisms of the DNA under irradiation (Banik et al., 2003).

The conventional thermal effects can possibly be manifested in the application of nonionizing radio frequency MW radiation which are used in industry, commerce, medicine, and in mobile phone-based communication technology. For example, experiments with low frequency (2450 MHz) MW irradiation on microbially contaminated toothbrushes show the disruption of biofilm structure and the data suggest strongly that the destruction of micro-organisms by MW is mainly due to a thermal effect. The same MW irradiation at a low-intensity level for a long exposure time (about 30 days), was however also found to cause DNA damaging effects in the form of breaking of strands leading to a mutation in the brains of Fischer rats (Deshmukh et al., 2013; Sarkar, 1994). The mechanism for this is not very clear, although several plausible processes have been suggested which are of either thermal or of magnetic origin.

On the other hand, the high-frequency MW radiation (in the range of several hundred gigahertz to tera-hertz) has a very clear possibility of undergoing resonance interaction with the biological systems. The reason behind this is that these frequencies fall well within the range of frequencies of conformational oscillations of both DNA and Protein molecules and also match quite well with those corresponding to the energies of the hydrogen bonds and Van-der-Waals inter-molecular interactions, as well. The direct experimental observation of resonance absorption of MW via interaction with the vibrational modes (phonons) presents within the above macromolecules turns out to be quite difficult and unreliable. However, the analysis of the absorption spectra from the scattering experiments involving a DNA film did establish the presence of resonance interaction between the MW radiation and the intra-molecular phonon modes of the DNA. On the other hand, exposure of real biological systems like *E. coli* cells to tera-hertz radiation of wavelengths ranging from 130–200 micrometers with a power of 1.4 Watt/sq.cm clearly induces changes in green fluorescent protein (GFP) fluorescence values, as seen in experiments (Copty et al., 2006). This is a direct manifestation of the mutation induced by the coupling of the DNA molecules to the high-frequency MW radiation, it is believed. This coupling too is expected to be mediated through the intra-DNA phonon modes (Demidova et al., 2012).

The phonon modes in these DNA molecules primarily arise due to the physical relocation and reorientation of the Hydrogen bonds forming the base pairs involving the two complementary strands, as well as from the bending, stretching, and torsional modes of the covalent bonds within various nucleotides themselves (Chaudhury, 2015; Golo, 2005). These distortional modes modify the magnitudes of both intra-strand and inter-strand couplings amongst the nucleotides significantly.

The theoretical understanding and treatments for the above processes leading to MW-DNA interaction, in particular, can be achieved with the help of various quantum models already developed for DNA, as introduced in the earlier section on TM. The hidden interaction between the phonons and the other electronic excitations represented as 'spins,' 'isospins,' and 'bosons,' occurring naturally in these theoretical approaches, can be handled readily with the standard condensed matter physics-based techniques [see the earlier sections on mutation]. This would further help in both qualitative and quantitative understanding of the full interaction between the MW radiation and the phonon modes of the DNA, including the resonance type of behavior. The theoretical calculations involve the modeling of frequency-dependent dielectric function for a DNA molecule, determining the phonon frequencies as well as the study of elastic dynamics of torsional acoustic and hydrogen-bond-stretch modes (Golo, 2005, 2014).

6.6   DNA MELTING

A sufficiently large thermal energy can cause a rupture of the chemical bonds in a DNA molecule. Consequently, at a very high temperature a DNA molecule 'melts,' i.e., the two complementary strands of DNA get totally delinked from each other. This is because of the breaking of the hydrogen bonds connecting the complementary bases. This phenomenon or process is known as the melting or denaturation of DNA. As the G-C bonds are stronger than the $A–T$ bonds, it implies that the (G+C) rich DNA configurations possess higher melting temperatures than those of (A+T) rich ones (Griffiths et al., 2002). Theoretically, the DNA melting temperature (T_m) is defined statistically, however. It is determined by the criterion that at least 50% of the hydrogen bonds have been broken at that lowest temperature, the magnitude of which can be obtained experimentally as well as by computational techniques (standard for empirical formulae of T_m). Experimentally, the DNA melting temperature is determined by noting the temperature at which the absorbance of the DNA solution at 260 nm corresponding to the UV range, exhibits an enhancement by 30 to 40% (Structural Biochemistry/Wikimedia book, 2000). Very often, the experimental value for T_m comes out to be around 80°C (Molecular Diagnostics, 2007).

As outlined in the above paragraph, the DNA melting is essentially a thermally driven process guided basically by classical statistical mechanics; nevertheless, the quantum effects can play important roles in deciding the magnitude of T_m. In particular, the expressions for mutational rate as well as for CR rate or base pair transition rate, as obtained by our quantum spin model in an earlier section, would determine quantitatively the role of pure quantum fluctuations on the melting of DNA. Intuitively, the quantum fluctuations are expected to lower the magnitude of T_m. This, however, can only be verified by detailed numerical calculations with our model at finite temperature involving quantum statistical mechanics, since mutation and CR processes produce opposite effects, as far as the breaking of the bonds between the complementary bases are concerned. It is also worthwhile mentioning that our quantum mechanical calcula- tional scheme has modeled the structure of DNA simply as a pair of linear strands with complementary base pairs facing each other, as is done in the well-known "De Gennes, Peyrard-Bishop" (GPB) description (De Gennes, 1969; Peyrard and Bishop, 1989). Our scheme, however, does not make use of any heuristically introduced inter-ionic pair potentials, *viz.* intra-base as well as nearest neighbor inter-base pairing potentials for complementary strands and intra-strand harmonic or anharmonic poten- tials involving nearest neighbor bases within a classical framework, as has been done in several other theoretical or computational approaches for the investigation of DNA melting from the chemical perspective (Khandelwal and Bhyrabhotia, 2010; Zhou et al., 2000). On the contrary, our theoretical methodology deals with only inter-strand spin-spin coupling parameter λ and intra-strand transverse field H, the basic ingredients for a description of DNA, present in our quantum modeling (as has been discussed in the earlier section on TM). The very quantum origin of tautomeric fluctuations can be utilized to directly investigate the melting or denaturation dynamics of DNA, by our procedure. Furthermore and very importantly, the DNA molecules being quasi-one-dimensional objects the role of fluctuations consisting of both quantum fluctuations, manifested explicitly in this chapter through the consequences of the calculations with the quantum models used in its descriptions, as well as pure statistical mechanical fluctuations, in the melting process can be quite non-trivial and substantial. This may very well have a large impact on the magnitude of the melting temperature for DNA. Therefore, this would definitely be a potential area for further theoretical research.

KEYWORDS

- **base-pairing rule**
- **complementarity restoration**
- **Holstein-Primakoff transformations**
- **repairing efficiency**
- **Schrieffer-Wolf transformation**
- **tautomeric mutation**

REFERENCES

Banik, S., Bandyopadhyay, S., & Ganguly, S., (2003). *Biosource Technology, 87*, 155.

Braun, J. V., & Muler, H. G., (1998). *Statistical Science, 13*(2), 142.

Buckingham, L. & Flaws, M. (2007). *Structural Biochemistry/Wikimedia.* Molecular diagnostics for clinical laboratory science, Lecture Notes by Buckingham, L., and Flaws, M. (2007).

Cairns, J., Overbaugh, J., & Millar, S., (1988). *Nature, 335*, 142.

Chaudhury, R., (2007). *Europhys. Lett., 79*, 18005.

Chaudhury, R., (2015). *Nanostructures, Mathematical Physics and Modeling, 12*(1), 85.

Chaudhury, R., Saraswathi, S. D. B., & Sasmol, A., (2019). (Under preparation).

Copty, A. B., Oz, Y. N., Barak, I., et al., (2006). *Biophys. J., 91*(4), 1413.

De Gennes, P. G., (1969). *Rep. Prog. Phys., 32*, 187.

Demidova, E. V., Goryachkovskaya, T. N., Malup, T. K., et al., (2012). *Bio-Electromagnetics.* doi: 10.1002/bem.21736.

Deshmukh, P. S., Megha, K., Banerjee, B. D., et al., (2013). *Toxicology International, 20*, 19–24.

Golo, V. L., & Volkov, Y. S., (2003). *Int. J. Mod. Phys. C., 14*(1), 133.

Golo, V. L., (2005). *Journal of Experimental and Theoretical Physics, 101*(2), 372.

Griffiths, A. J. F., Lewontin, R. C., Gelbart, W. M., & Miller, J. H., (2002). *Modern Genetic Analysis: Integrating Genes and Genomes.* Chapter 2, W. H. Freeman and Company, New York.

Gujjari, S. K., Gujjari, A. K., Patel, P. V., & Shubhashini, P. V., (2011). *J. Int. Soc. of Preventive and Community Dentistry, 1*(1), 20.

Karlin, S., (2005). *Proceedings of National Academy of Sciences (USA), 102*(38), 13355.

Karlin, S., Ost, F., & Blaisdell, B. E., (1989). In:Waterman, M. S., (ed.), *Mathematical Methods for DNA Sequences* (pp. 133–158). CRC Press, Boca Raton, Florida.

Katabeh, Q. D., Aroza, J. M., Lopera, J. F. G., & Navarro, D. B., (2015). *Entropy, 17,* 7996.

Khandelwal, G., & Bhyrabhotia, J., (2010). *PloS One, 5*(8), e12433. doi: 10.1371/journal.pone.0012433.

Lin, J., (1999). *IEEE Transactions on Information Theory, 37,* 1.

McFadden, J., (2000). *Quantum Evolution: Life in the Multiverse.* Flamingo, London.

Peyrard, M., & Bishop, A. R., (1989). *Phys. Rev. Lett., 62,* 2755.

Sarkar, S., (1994). *In Micro Wave News* (Vol. 14, No. 6). A report on non-ionizing radiation.

Shannon, C. E., (1948). *The Bell System Technical Journal, 27,* 379.

Som, A., Sahoo, S., Mukhopadhyay, I., Chakrabarti, J., & Chaudhury, R., (2003). *Europhys. Lett., 62,* 271.

Woolard, D. L., Globus, T. R., Gelmont, B. L., et al., (2002). *Phys. Rev. E., 65,* 051903.

Zaburdaev, V. Y., (2006). *Journal of Statistical Physics, 123,* 871.

Zhou, H., Zhang, Y., & Yang, Z. (2001). Theoretical and computational treatments of DNA and RNA molecules (2000); *Phys. Rev. Lett., 86,* 356.

Index

For Product Safety Concerns and Information please contact our EU
representative GPSR@taylorandfrancis.com
Taylor & Francis Verlag GmbH, Kaufingerstraße 24, 80331 München, Germany

www.ingramcontent.com/pod-product-compliance
Lightning Source LLC
Chambersburg PA
CBHW052013230326
41598CB00078B/3213